BUJUTSU
AN ILLUSTRATED GUIDE

A Translation of:

兵法要務：柔術剣棒図解秘訣

Fundamentals of Military Strategy:
Secrets of Jujutsu, Kenjutsu & Bojutsu
An Illustrated Guide

井口松之助著
By Inoguchi Matsunosuke
Published 1887

Translated by Eric Shahan

Translator's introduction

Bujutsu is the first illustrated guide to martial arts published following the 1868 Meiji Restoration which ended the feudal system under a Shogun and restored all lands to the emperor of Japan. With this change, the Samurai class disappeared, as did the source of their livelihood. Samurai were forced to create businesses in order to make a living. This book, *Bujutsu,* was an early effort to make the transition from an employee of the government to a citizen earning a living selling martial arts instruction.

The author, Inoguchi Matsunosuke (birth and death dates unknown, active until 1930s) introduces Jujutsu, sword, short sword, staff and half-staff techniques in addition to marital arts philosophy harnessing supernatural powers.

In the years following the release of this book, which went through multiple reprints, dozens of other authors followed suit and published illustrated guides to various martial arts. The success of this book may have been the catalyst that started a whole new genre.

Though there is little information about the life of Inoguchi Matsunosuke, who also published under the name Noguchi Yoshitame, he authored several books about martial arts up until the 1930s. Given his extensive experience with martial arts, he was likely born into a Samurai family that served as martial arts instructors for one of the former domains of Japan.

Regarding this Translation

Japan dramatically simplified the official writing style following the second World War, so any books from the late 19th and early 20th centuries are written in an older style that can be quite difficult to read (even for Japanese people.)

Many books from that period:
- Contain Kanji that are written in an older, more complex style
- Contain text that is written with Kanji and Katakana instead of Kanji and Hiragana
- Limited punctuation

The biggest obstacles to making a clear translation of the martial arts techniques were the transitions from one movement to the next. To facilitate this, some information was taken from later books that contain the same techniques as *Bujutsu*. The books are *An Illustrated Guide to the Inner Mysteries of Tenjin Shinyo School Jujutsu* (1893) and *Tenjin School Jujutsu* (1926.)

Some illustrations from these books will be included to help clarify the techniuqes. They will be differentiated based on their date of publication.

The "1893 Version,"
- *Tenjin Shinyo Ryu Jujutsu Gokui Kyoju Zukai Jujutsu*
  天神真楊流柔術極意教授図解
  Illustrated Guide to the Inner Mysteries of Tenjin Shinyo School Jujutsu
  By Yoshida Chiharu 吉田千春 and Iso Mataemon 磯又右衛門
  Published 1893

The "1926 Version,"
- *Tenjin Tesshin Ryu Jujutsu Kata Gokui Hiden Zukai: Hokoku Kan Tecchu Ryu*
  天神鉄真流柔術型極意秘伝図解：報国館鉄仲流
  Illustrated Guide to the Inner Secrets of Tenjin School Jujutsu Kata: Hokoku Hall Tecchu School
  By Kushi Niju 大串仁十
  Published 1926

INOGUCHI MATSUNOSUKE · ERIC SHAHAN

大法螢雪芳

正四位伯爵勝安房君題字　魁真

正四位子爵山岡鐵太郎君序　棲に蔵

榊原健吉君校閲廿後枝

松涎舎編輯版

版所兼攝

柔術劍棒圖解秘訣

獨聲古

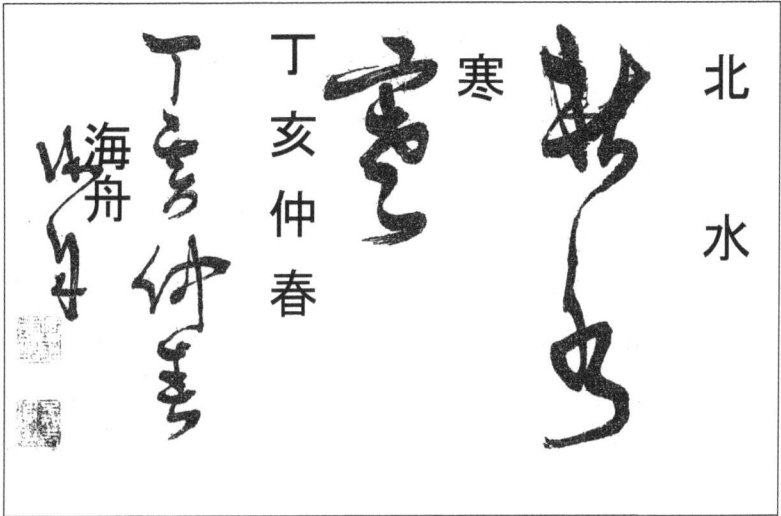

The calligraphy on the previous pages:

北水寒
丁亥仲春
海舟

The frigid water of the north in winter
February 1887 Kaishu[1]

---

[1] This calligraphy is by Katsu Yasuyoshi 勝安芳 (1823 ~ 1899,) who went by the name Katsu Kaishu. He was an engineer, statesman and military commander.

The phrase describes the utter coldness of water in the depths of winter, which is thought to have an almost divine power. In particular *Kankyu no Sui* 寒九の水 "water on the 9th day after midwinter," is said to be the clearest and purest water. Thus, makers of Sake, soy sauce and Miso collect water on this day to improve the taste and durability of their products. Further, drinking this water is thought to reinvigorate the body.

華子岡寺

阮

榮悴榮

莫道滋月

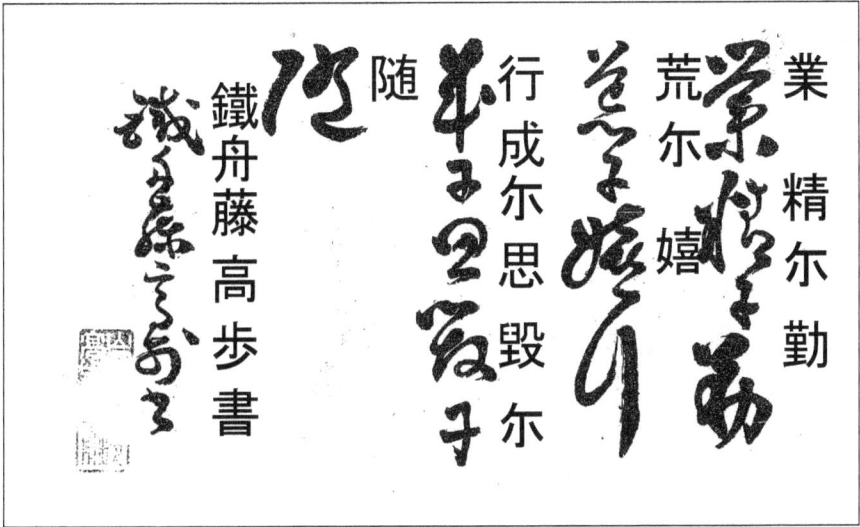

The calligraphy on the previous pages:

業精尔勤荒尔嬉
行成尔思毀尔随
鐵舟藤高歩書

If you continue to study diligently you can expect to meet great success. However, if you are lax in your study, expect your life to fall into disarray.

By Yamaoka Fuji Takayuki[2]

---

[2] This is a line by the Confucian Scholar Han Yu 韓愈 (768 ~ 824.) It was written by Yamaoka Fuji Takayuki Tesshu 山岡鐵舟 (1836 ~1888) a swordsman, calligrapher and heavy drinker.

叙言

一 此書ハ兵法ノ要務柔劍棒三術圖解秘決一名獨誓古ト
題ス尚ホ現今ノ専用タル早繩鎖ノ二類ヲ加ヘ合セテ
五法數手盡ク書畫ヲ以テ之ヲ詳細ニス專ラ讀者ヲシ
テ安カランコヲ旨トス

一 柔術ハ受方ヲ甲者ト為シ捕方ヲ乙者ト為ス手術表裏
ノ解書ノ尽サ丶ルハ圖畫ヲ以テ辨シ圖畫ノ及ハサル
ハ又書ニ讓ル讀者第一圖ヨリ第二圖第三圖ト係連シ
テ視ルベシ仕組手解拾二組初段居捕立合都合二拾組
投捨二拾組口傳秘決多シト雖モ大緊ネ之ヲ洩ス事ナ
シ其書畫ニ尽シ難キ物ハ目録ノミヲ挙テ粗畧スル所
アリ

Preface

Bujutsu, the full title of which is *The Fundamentals of Military Strategy: An Illustrated Guide to the Secrets of Jujutsu, Kenjutsu and Bojutsu,* should be considered a guide to solo training in the three aforementioned martial arts. In addition, by including the specialist techniques *Haya Nawa* (how to quickly tie rope handcuffs) and *Kusari* (self-defense using a weighted chain) a total of five martial arts are introduced in this book. For the benefit of the reader, several techniques from each art will be described in detail with illustrations.

For the Jujutsu techniques the *Uke-Kata*, the one receiving the technique, will be labeled as *Koh*, former and the *Tori-Kata*, the one doing the technique, will be labeled as *Otsu*, latter. The best way to understand all facets of a technique is to first study the illustrations and then, if any points are unclear, refer to the text. The reader should focus on how Illustration 1, Illustration 2 and Illustration 3 of each technique are connected to one another.

The Jujutsu section will include the following:
    Freeing Your Hands   – 12 Techniques
    First Level Seated   – 20 Techniques
    Sacrifice Throws   – 20 Techniques

While there are many *Kuden Hiketsu*, secrets only transmitted verbally from master to student, due to the illustrated descriptions, the reader will have no problem understanding the fundamentals of each technique. That being said, there are a few techniques that are difficult to present through pictures and will therefore be abbreviated with only the name listed.

一擊劍ハ日常ノ仕合ハ變化無窮ニシテ畫圖ニ筆シ難シ故

二其一二ヲ擧ルノミ陰陽ノ構ヘ擊突拂應及ヒ短刀ノ

用法ホハ悉ク書ヲ以テ之ヲ明瞭ニシ尚ホ先師看破ス

ル所ノ論言ハ具サニ之ヲ記載ス

一棒及ヒ早繩鎖ノ用法總ク圖畫ニ依ル之ヲ弁明ス六尺

捧拿一番ヨリ第八圖迄三尺棒第一圖ヨリ第六圖迄順

順序ヲ以テ之ヲ明カス其他九字法十字法ノ秘決或ハ

理哥或ハ土段卷藁ノ製方迄悉ク記載シ以テ武術ヲ修

得スル諸君ノ覽ニ備フ請フ之ヲ實際ニ試ミテ其虛ナ

ラサルコヲ撿セヨ

編者識

Regarding the section on *Gekken*, Japanese style fencing duels, these techniques rely on the practitioner being able to respond freely and instantly to any attack your opponent makes. Therefore, as it is difficult to capture this in pictures, each technique will consist of just two illustrations. In the techniques you will be in either Yin or Yang stance and be attacking with, or responding to, cuts, stabs and sweeping attacks. In addition, a clear explanation of Tanto, short sword, techniques will also be introduced. The chapter will conclude with some astute observations by a master instructor.

Bojutsu, six-foot wooden staff fighting, as well as *Haya Nawa*, how to quickly tie rope handcuffs, and *Kusari*, self-defense using a weighted chain, will be explained through illustrations. *Bojutsu* will be introduced from Illustration 1 to Illustration 8, while *Sanjakubo*, three-foot half-staff, will be introduced from Illustration 1 to Illustration 6. By following the illustrations in order, the reader will clearly understand how this weapon is used.

In addition, the secrets of *Kuji*, the nine seals, and *Juji*, the tenth seal, will be introduced along with the proper incantations. For training purposes, instructions for making both an earthen mound and a rice straw bale for sword practice have been included. This is the ideal book for those interested in training Bujutsu, martial arts, and by trying the techniques out in real life, you will find they do not deceive you.

-Editor

兵法柔術剣棒圖解秘訣總目

BUJUTSU
Fundamentals of Military Strategy:
Illustrated Guide to the Secrets of Jujutsu, Kenjutsu & Bojutsu

TABLE OF CONTENTS

仕合ニ八箇條ノ法アル論

心気力一致ノ事 ○ 七知ノ教ノ事

柔術ノ部

手解十二手図解

名称 ○兎拳 ○振解 ○逆手 ○逆指 ○片胸捕 ○両胸捕 ○小手返 ○両手返 ○気捕 ○天倒 ○扱取 ○打手

初段居捕十手図解

全 ○真ノ位 ○添捕 ○御前捕 ○袖車 ○乗違 ○抜身目付 ○鎧返 ○両手捕 ○壁添 ○後捕

同立合 十手図解

全 ○行違 ○突掛 ○引落 ○両胸捕 ○連拍子 ○友車 ○絹潜 ○襟投 ○手髪捕 ○後捕

同投捨図解

全 ○襟投 ○小具足 ○撞木倒 ○朽木倒 ○腰車 ○横車 ○片胸捕 ○手髪捕 ○腰挟捨 ○独鈷 ○小手返 ○引落 ○手繰 ○下藤 ○矢筈 ○両手捕 ○両柄捕 ○後捕 印ノ四手ハ口傳多ク危險ナレバ畧ス

乱捕 土水

全 ○横捕 ○小手ぎ ○襟捕 ○突込 ○胴〆 ○組合突込 ○捨身 ○腕シギ ○肌我 ○腰投 ○捨身投 ○背負投

Jujutsu

Tehodoki: Freeing Your Hands
12 Illustrated Techniques

1.	鬼拳	Onigoshi/ Oni Kobushi:	Devil's Fist
2.	振解	Furi-hotoki:	Shake Free
3.	逆手	Gyaku Te:	Reverse Grip
4.	逆指	Gyaku Yubi:	Bending Fingers Back
5.	片胸捕	Kata Mune Dori:	One-Handed Chest Grab
6.	両胸捕	Ryomune Dori:	2-Handed Chest Grab
7.	小手返	Kote Gaeshi:	Bending Back the Hand
8.	両手返	Ryote Kaeshi:	Two-Handed Reverse
9.	気捕	Kidori:	Seizing the Chance
10.	天倒	Tento:	Top of the Head
11.	扱捕	Mogi Dori:	Plucking Away
12.	打手	Uchi Te:	Striking Hand

Shodan Idori
Seated First Level Techniques
Ten Illustrated Techniques

1.	真之位	Shin no Kurai:	True Stance
2.	添捕	Soe-dori:	Alongside Seizure
3.	御前捕	Gozen Dori:	Before Royalty Technique
4.	袖車	Sode Guruma:	Sleeve Wheel
5.	飛違	Tobi Chigai:	Leaping In and Attacking
6.	抜身目付	Nukimi Metsuke:	Locking Eyes and Drawing
7.	鐺返	Kojiri Gaeshi:	Reversing the Scabbard Cap
8.	両手捕	Ryo-te Dori:	Two-Handed Attack
9.	壁添	Kabe Soi:	Pressed Against a Wall
10.	後捕	Ushiro Dori:	Attacked from Behind

Shodan Tachiai Jutte
First Stage : Standing Techniques
Ten Ilustrated Techniques

1. 行違　Yuki Chigai　　Crossing Paths
2. 突掛　Tsuki Kake　　Attacked With a Punch
3. 引落　Hiki Otoshi　　Pulled Down
4. 兩胸捕　Ryomune Dori　Two Handed Chest Grab
5. 連拍子　Renhyoshi　　Joined in the Same Rhythm
6. 友車　Tomoguruma　　Roll Together
7. 絹潜　Kinu Katsugi　Dropping and Loading Like Silk
8. 襟投　Eri Nage　　Collar Throw
9. 手髪捕　Tabusa Dori　Someone Grabbing Your Hair
10. 後捕　Ushiro Dori　Grabbed from Behind

Nagesute
Throwing

Twenty Illustrated Techniques

1.	撞木	Shumoku	Piercing Tree
2.	苅捨	Karisute	Like a Scythe Through Grass
3.	朽木倒	Kuchiki Taoshi	Toppling a Decayed Tree
4.	腰車	Koshi Guruma	Tossing Like a Wheel
5.	横車	Yoko Guruma	Side Toss Like a Wheel #1
6.	片胸捕	Katamuna Dori	One-Hand Chest
7.	手髪捕	Tafusa Dori	Seized by the Hair Technique
8.	小具足	Kogusoku	Fighting Against Short Weapons
9.	腰苅捨	Koshi Kari Sute	Hip Sweep and Dispose
10.	獨鈷	Dokko	Vajra
11.	小手返	Kotegaeshi	Reversing the Wrists
12.	引落	Shiki Otoshi	Pulling Down
13.	手繰	Taguri	Controlling with the Hands
14.	捨身	Sutemi	Sacrifice Throw
15.	下り藤	Kudari Fuji	Descending Wisteria
16.	腕縅	Ude Karami	Arm Tangle
17.	矢筈	Yahazu	Nock of an Arrow
18.	両手捕	Ryote Dori	Two-Handed Capture
19.	両柄捕	Ryotsuka Dori	Two-Handed Handle Capture
20.	後捕	Ushiro Dori	Seized From Behind

撃劍ノ部

Randori
Free Sparring
Twelve Illustrated Techniques

1.	*Nyu Tori*	Chest Grab
2.	*Eri Tori*	Seizing the Collar
3.	*Kote Shigi*	Lower Arm Twist
4.	*Tsuki Komi*	Piercing In
5.	*Dojime*	Waist Strangle
6.	*Sutemi*	Sacrifice Throw
7.	*Tsuki Komi*	Piercing In
8.	*Ude Shigi*	Arm Lock
9.	*Hadaga Tori*	Naked Self-Choke
10.	*Sutemi Nage Kamae*	Preparing for Sacrifice Throw Stance
11.	*Koshi Nage*	Hip Throw
12.	*Seoi Nage*	Shoulder Throw

Gekken – Japanese Fencing Techniques

- o Why you should study the short sword
- o Description and Illustration of Makiwara and Dotan
- o Toda School Weighted Chain and Illustration
- o The Rule of Near and Far
- o Japanese Fencing Techniques and Illustrations
- o Short Sword Techniques
- o Six-foot Staff Techniques
- o Sword Versus Six-foot Staff Techniques
- o Half Staff Techniques
- o Striking Chart
- o Illustrated Guide to Fast Tie
- o How to Tie a Fast Tie
- o How to Tie a Short Tie
- o An Illustrated Guide to Kuji
- o An Explanation of Juji

End of Table of Contents

○柔術劒術棒之総論

皇國往古ヨリ兵法ノ要務タルハ何ヤ曰ク弓馬劒槍柔ノ
五種ナリ星移リ世代リ百物変遷スルト雖モ日常護身ノ
要用無カル可カラズ夫禽獣ニ牙角爪アリ史ニ剋アル是
皆護身ノ具ナリ人トシテ尚モ不虞ノ備ナキ時ハ鳥獣ニ
若サルナリ然リ而レテ目今衆庶護身ノ具トシテ學フ可
キハ何ヤ曰ク劒術柔術曰ク棒ノ類ナリ其故如何トナ
レハ劒ハ我邦神代ニ其制ヲ速シ名劒ノ威德萬國ニ卓越
スルハ自他信ヲ置ク所ナリ故ニ歴代ノ帝王大ニ劒德ヲ
尊ヒ常ニ王躰ヲ離チ王ハサル所ナリ其萬國ニ冠タル
劒ノ獨吾人世ニ傳ハリタルハ之ヲ使用スルノ術又無カ
ルヘカラズ是故ニ先師柔術法ヲ練磨シ其妙手ヲ極ムル

Overview of Jujutsu, Kenjutsu and Bojutsu

Now that the imperial family once again rules Japan, do we not have a duty to train martial arts? The five martial arts often mentioned are: archery, equestrian, sword, spear and Jujutsu. Clearly, in these times, when everyone is at the mercy of the vicissitudes of the age, you should not go about without a knowledge of the fundamentals of self-defense.

Birds and beasts possess fangs, horns and claws while insects can sting. These are all tools that creatures use to defend themselves, unfortunately humans are not equipped with any such weapons. At times it seems we aren't as fortunate as the animal kingdom. So what tools are the members of the public equipping themselves with for self-defense nowadays? The answer is Kenjutsu. The answer is Jujutsu. The answer is also various types of pole arms.

Looking back to the age of the gods, the first weapon used to bring order to Japan was the sword. Great men in our past used swords of great renown to lead their domains to success and defend its borders, while understanding that their neighbors intended the same.

Thus, for generation after generation the great rulers of Japan respected the virtuous authority of famous swords and kept them at their sides. The crowning achievements of each domain of Japan were the sword makers and those sword makers have passed on the tradition to us. The use of these weapons is not something we should lose.

著各其流派ヲ立テ以テ之ヲ後人ニ傳フ其流得幾種ナレ

㆓尤古クヨリ盛ニ世ニ行レタルハ神蔭流后チ直心影

流ニ改ム一刀流無念流一傳流柳生流柳剛流等ナリ○

桑術ハ古クヨリ其妙ヲ傳ヘ治亂ニ亙リ學ハサル可カラ

サルノ要術ナリ其術タル所謂柔能ク剛ヲ制スルノ道ニ

シテ我ニ勝レルカ者ト雖ㆳ容易ク之ヲ捕挫ギ已ヲ全フ

シテ必勝ヲ繰ツノ術ナリ夫カハ限アリ術ハ限リナシ謂

ハンヤ死活ヲ自在ニシ其奥ヲ極ムルニ至ッテハ夭死者

ノ克ク蘇活セシムルアリ嗚呼該世ノ衆實ニ学バスンバ

有ル可カラブ其流派ノ世ニ著キモノハ揚心流神道流真

揚流起倒流等ナリ流義ニ因リ教法大同小異アリト雖ㆳ

其奥妙ニ至ッテハ一ナリ○搶術ハ其長器タルヲ以テ

These sword practitioners also practiced Jujutsu. The Jujutsu practitioners refined their skills until they captured the essence of the art. Each of these early masters of Jujutsu went on to set up a school that taught their own style, which was later passed on to their successors.

There are many different names for these Jujutsu schools, the oldest of which is Shin Kage Ryu, Divine Shadow School, which later became Jikishin Kage Ryu, Direct Mind Shadow School. This school spread all over Japan. There are also sword fighting schools like Itto School, Munen School, Ichiden School, Yagyu School and Ryuko School..

Jujutsu

From days long past the essence of Jujutsu has been transmitted through both times of war and times of peace. As Jujutsu is a fundamental art, it was unconscionable for it not to be part of the curriculum that Samurai studied. The underlying principle of Jujutsu is that your soft and the flexible technique can control a strong and rigid opponent, leading to your victory. The art of Jujutsu allows you to easily break your opponent's technique even if he is of superior strength and invariably defeat him. As the saying goes,

While there is a limit to your physical strength there is no limit to the degree to which you can refine your Jujutsu technique.

This means that if you develop an understanding of the inner mysteries of Jujutsu, then you will be able to restore unconscious people to life giving you complete control over life and death.[3] Alas, unfortunately many people today do not seek to learn this art.

Some of the well-known schools of Jujutsu are the Yoshin School, the Shinto School, the Shinyo School, the Kito School as well as others. If you look at these schools closely you will find the way each school teaches is broadly similar and varies only in the details. Further, once you have developed an understanding of this art, you will find all schools of Jujutsu are based on the same fundamental principles.

---

[3] This is referring to using vital points on the body. Striking them can result in death while using them effectively can restore an unconscious or drowned person to life.

今世学ブ者稀ナリ依テ之ニ代ユルニ棒ヲ以テス棒ハ八丈

ケ六尺半棒ハ三尺ニシテ護身ニ切用アリ荒木流抔手術

ククミ
巧妙ナリ

○柔剱棒学シテ切アル辨

前ニ謂フ処ノ三術ハ諜世ニ於テ学シテ大ニ切用アリ先

一二ヲ去ハン譬ヘハ常ニ路頭ヲ往來スルノ際醉狂人又

ハ白痴者抔ノ乱暴ニ逢ヒ或ハ嫌行シテ原野ニ至リ惡漢

盗兒ノ為ニ不慮ノ災害ヲ蒙リ或ハ夜陰強窃ノ盗難ニ罹

ルノ時庸人ハ大ニ心周章シテ何ノ分別モ出サルモノナ

リ常ニ護身ノ術ヲ学ヒ深慮活潑ノ精心ヲ養フモノハ事

ニ臨ンテ動搖セズ己ヲ全フシテ難ヲ避ルコ古今其例少

ナカラス

Sojutsu

Sojutsu, or Spear fighting, is about training with pole arms. These days it is rare to find someone who is interested in training spear, thus many people train Bo, wooden staff, instead. A wooden staff is typically *Rokushaku*, 6 feet long, and *Hanbo*, Half Bo, are *Sanshaku*, 3 feet, long. This art is effective for self-defense and schools like the Araki School are known for their refined wooden staff techniques.

An argument for studying Jujutsu, Kenjutsu and Bojutsu

As was previously mentioned, the general populous will find that there is a great benefit to learning the Three Arts: Jujutsu, Kenjutsu and Bojutsu.

With regards to the first two, think of a situation where you are walking along a road, maybe as part of your regular commute, and you encounter a deranged individual, who is wandering about. Or perhaps you encounter an idiot or some other violent person. Or perhaps you are on vacation and travelling through the countryside and you encounter bad men or thieves who make untoward demands on you, leading to disaster. Or it is late at night and you find yourself at the hands of a robber. In such cases, many average folks become frightened and are rendered unable think of what to do. This is because they have not done the prudent thing and made training self-defense part of their everyday lives, thus do not have the mental fortitude necessary to respond.

Looking back across history, you will find innumerable examples of people remaining steadfast in such situations, thereby allowing them to emerge unscathed.

○初學心得

前三術ヲ学フ者其最初ニ於テ尤心得ベキ一アリ先ツ劍

法ヨリ去ハン古ヘ劍道ヲ傳習スルノ法ハ現在ト異ナリ

ニ依ラズ洗練ノ渚破發明スル所ノ構ヘ上段下段星眼平

星眼八相遮ノ類或ハ撃チ突キ或ハ拂ヒ應スルノ妙手ヲ

集メ彩チヲ造リ木太刀ヲ以テ習練シ驚懼疑惑ノ怯心ヲ

去リ太刀筋ヲ正シクシ氣息ヲ冨マシメ撃チテハ必ス切レ

突ケハ必ス貫リ敵ノ変化ニ應シテ捨身護身ノ構ヘヲナ

スコヲ教ヘタリ稍上達スルニ及ンデ其業ノ巧拙ヲ試ン

ガ為メ稀ニ木太刀ヲ以テ頑負ノ勝負ヲ決セシナリ後世面

小手胴抔ノ具ヲ自製シ竹刀ヲ造リテ恒ニ勝頑優劣ヲ争

フヲ以テ誓古ト云フ劍法ヲ学ハントスル者撃チ突キ勝

Lesson for Those Beginning Training

There are important things to learn when beginning training the above mentioned Three Arts: Jujutsu, Kenjutsu and Bojutsu.

First of all, when studying Kendo, the Way of the Sword, understand that is derived from old-style Kenpo, the Law of the Sword. The methodology of Law of the Sword differs from what is taught nowadays. The sword stances taught today were developed through clever insights by early masters of the art. They are,

Jodan	–	Upper Stance
Gedan	–	Lower Stance
Seigan	–	Middle Stance
Hira Seigan	–	Open Middle Stance
Hasso	–	Middle Upper Stance

There are other stances as well. Sword techniques include,

Uchi	–	Strike
Tsuki	–	Stab
Harai	–	Sweep
Ukeru	–	Block (responding to your opponent)

Various combinations of the above stances, attacks and defenses were compiled into Kata, series of prescribed offensive and defensive moves. Students would train with *Kidachi*, a Wooden Long Sword, to develop a strong mind that can remain calm when facing an opponent, while at the same time eliminating fear and doubt. Students were taught how to swing their sword with correct *Tachi Suji*, Cutting Lines, as well as how to maximize effective breathing while increasing their martial spirit.

Following this program meant you would be able strike in a way that always resulted in an effective cut or a thrust that will invariably penetrate your opponent.

This book will also include information on *Sutemi Goshin Kamae*, Sacrifice and Self-Defense Stances, meaning you will move as close as possible to your opponent risking your own life.

Up through the 1600s, once you reached expert level, you would take advantage of rare opportunities to enter a duel in order to test the relative strength or weakness of your technique. Though such duels are fought with wooden swords, it could well be a life-or-death contest.

員ヲ争フ物トノミ思ハズ師ニ依リテ其傳來ノ形ヲ學ヒ

之ヲ日々試合ノ上ニ施シ修業セハ早ク妙所ニ至ルヿヲ

得ン今モ天眞流抔ハ木太刀ヲ以形ヲ學ヒテ試合ト云ヿ

ハ爲サヽルナリ　柔術棒ハ巻中悉ク其形ヲ圖シ釈ヲ

詳細ニシ諸生ヲシテ自ラ其妙手ヲ得セシム柔術ノ形ヲ

實際ニ施シ試ムルヲ乱舐ト云フ無双新流抔ハ乱取ヲ以

テ日々ノ傳務トス　○今壯年初テ武術ノ門ニ入リカメ

テ其奧義ヲ極メントスルノ志操ハ實ニ嘉スヘク賞スヘキ

事ナリ然ルニ未ダ幾バクナラズシテ稍モスレハ窃ニ驕

心ヲ生シ吾ハ某ノ門人ナリ吾ハ何流ノ學士ナリ願ハク

ハ彼ヲ斬ケテ吾ガ掌中ヲ試ミン願ハクハ彼ヲ投ケテ我

カ事ヲ顯サント好ンテ災害ヲ招ク者アリ往昔ヨリ之ガ

In the mid-1700s, people began making protective armor to cover the head, hands and body.[4] Training swords were constructed out of bamboo, thus sword training in that era consisted of students in armor competing see who had superior skill.

However, sword practitioners that think the only way to train is to cut and thrust at opponents until achieving victory are mistaken. Many sword masters emphasize studying Kata, a series of prescribed offensive and defensive moves, before entering duels. This method enables you to master the fundamentals rapidly.

Schools such as the Tenjin School and others teach prescribed patterns to students using wooden long swords and do not engage in duels.

The Jujutsu and staff fighting techniques will be introduced with detailed illustrated descriptions, which will enable learners to master the fine points. To actually test yourself in Jujutsu, you must do *Randori,* free sparring. Schools such as the Muso Shin School and others make daily *Randori* training part of their curriculum.

○ Youth of Today

Nowadays, many young men are joining Dojo and working to develop an understanding of the inner mysteries of martial arts. This is a positive trend that is worthy of praise. However, some of these youth do no more than take a quick peek at what is happening in a Dojo and then immediately become overconfident, going about and saying, "I am a student of such-and-such school, so if you seek a duel, I will cut you down!" Or, "If you seek to challenge me, I will handle you with ease and throw you down!" More than a few people that take this route end up getting themselves into a situation that ends in disaster. Proving the aphorism, *A little learning in martial arts can lead to a great injury.*[5]

---

[4] Naganuma Kunisato 長沼国郷 (1688~1767) of the Jikishin Kage School of sword is often credited with developing early Kendo training armor and weapons.

[5] *Namabyouho wa O-kega no Moto* 生兵法は大怪我のもと A little learning in the arts of a warrior is a dangerous thing

為ニ身ヲ過ッ者獲少ナラス諺ニ曰フ生兵法大疵ノ元ナ

ルモノ是ナリ苟モ武術ノ門ニ入ルモノハ常ニ容貌温和

ヲ旨トシ事ニ臨ンテ活發敢勇鋭モ後レヲ取ヲサルヲ旨

トモ謂ヘキナリ

○度數ヲ重ヌル辨

修業最初ニ於テ自他流義ノ巧拙ヲ論シ或ハ已ガオカヲ

頼ンテ志高上ニ亘リ誓古ノ度數減少ナル者ハ到底上達

為シ難シカメテ我ヨリ上手ノ對手ニ順ヒ一向度數ヲ量

スルヲ主トスベシ俚歌ニ曰ク「ふ藝用と人ハいふことを彎

らせよ藝用そうりいいりていのるへき」又「藝藝いことを思

ふへにならふみちの邉の露の命のひをときゆるとを

るもひはに理のみ張してやゝとゝまぬる又「學へたゝゆ

That being said, when you join a martial arts Dojo, you will find that your overall appearance becomes calmer and gentler, while at the same time you feel more energetic and braver. You will develop into a person who is the embodiment of a solid gentleman.

⭕ An Argument for Consistent Training

After beginning training there is a tendency for people to compare the relative strengths and weaknesses of other schools when compared to their own or to train relying wholly on their physical strength in order to advance quickly. This leads to students training less frequently, which in turn actually makes it quite difficult to become an expert. The way to prepare yourself to face a more skilled opponent is to increase the number of times you train.   There is a folk song that goes,

Even if you think you are clumsy, you should practice until you become dexterous

Also,

Being lax in physically training martial arts and instead spending that time just on theory only leads to a decline in your overall skill

Also,

Study, but know that what you learned will be gone by morning like the dew on the side of the road

○撃劍執握ノ法

撃劍ニ於テハ柄ヲ執握スル法ニ注意スベキ手ノ内堅キ
ニ過クレハ修枝遲濶シ死物トナリテ活用自在ヲ失フナ
リ故ニ執握ノ法ハ左右共小指ヲ少シ締メ無名指沖指ハ
夫ヨリ輕ク込添ル計リニシ事ヲ施スノ際ダ
ハリト締メ打込ミ突キ出スナリ故ニ執握ノ法ヲ論ス
掌中雞卵ヲ握ルニ譬ヘ又小指ノ締リヲ茶布絞ナゾト譬
ヘヲトリ教ヘタリ直心流ニテハ木刀ノ柄ヲ執握スルニ
左手ノ小指ハ柄ノ頭ヲ外ス心得ニ教ヘタリ斯テハ眞
自在ニスル為トリ兎角左右ノ指ノ締リ方ニテ掌中クル
ヒ刀劍ノ又ハ當ラズシテ平ニテ撃ツモノナリ斯テハ眞
劍ニ於テ大ナル不覺ヲ取ル事必定ナリ常ニ太刀筋ヲ吟

○ How to Grip a Fencing Sword

When holding the sword in *Gekken*, Japanese fencing, or Kenjutsu, traditional Japanese sword fighting, you should be careful not to grip the handle too tightly.[6] If your grip is too tight, you will have difficultly employing techniques since your movements will be slow or hindered by the strength in your hands. This means you will have lost the ability to react freely to your opponent, which makes you an easy target.

The proper way to hold a sword is to first grip lightly with the little fingers of both hands. Next, curl your ring and middle fingers lightly around the handle. As for your index finger, it should simply be tucked alongside the others. When cutting or thrusting you should squeeze all your fingers together tightly. Thus, it is often said that you should hold the handle of your sword as if you are wrapping your hands around an egg. Another example is to imagine you are wringing out the cloth you clean teacups with. This means pulling down with your left little finger while squeezing forward with your right hand.

Students of the Jikishin School of Sword are taught to hold the handle of their wooden training sword so that the little finger of their left hand is curled outside the pommel. This allows a sword fighter to manipulate and move the sword freely. In conclusion, you need to be aware of how your fingers grip your sword. Gripping too lightly can also mean the handle twists in your hands, meaning instead of cutting your opponent with the blade, you only hit your opponent with the side of your sword. Such a blunder could be fatal if you were in a duel with a real sword.

---

[6] Generally speaking, *Gekken* focuses more on free sparring, while *Kenjutsu* focuses more on Kata training.

味シ土殺卷藁才ニテ手ノ裡ヲ試ムベシ尚ホ土段卷藁ノ
コヲ後段ニ辨センノミ

○食量ノ戒

武術ヲ学フ者ハ食量ヲ減スベシ目今文學ノ諸生運動ノ
足ラサルヲ以テ大ニ食量ニ注意ス是レ消化ノ互シカラサ
ル而已ニ非ラス食過ルトキハ氣息短促動作心ニ任セス若
悶忽チ至ル是故ニ武術ノ先師當テ暴食ヲ戒メタリ殊ニ
試合ニ望マントスル前ハ努メテ食量ヲ減少シ運轉自在
ヲ計ルヘシ

○對手ヲ擇ムヲ戒ム

何藝ニヨラス練習スルニ相手ノ好キ嫌ヒヲ爲ル者アリ
之レ大ヒニ上達ヲ妨クルノ第一ナリ如何ナル難物刀者

○ A Caution Regarding Your Diet

If you are studying martial arts, then you should reduce the amount of food you eat. Not only do literary people and others not get enough exercise these days, they also are not paying attention to the volume of food they consuming. This doesn't just affect your digestive system, eating too much can result in shortness of breath which, in turn, also puts pressure on your heart which could cause you to pass out. Thus, martial arts teachers have long cautioned their students against gorging themselves at meals.[7] In particular, before a duel, you should make every effort to reduce the amounts you eat in order to give yourself total freedom of movement.

○ A Caution Against Being Too Selective of Your Opponent

No matter what martial art you are doing, when training there will be some training partners you like to train with and some that you do not. However, this way of thinking is the prime barrier to becoming an expert. It doesn't matter whether your resistance is due to the fact that a certain opponent is difficult to work with, or because he is particularly powerful, you need to make a point of training with him. The act of thinking of different ways to handle such an opponent as you train will, conversely, be the key to your success.

---

[7] The author uses the word *Baku-gui* 爆食い eat explosively.

抔ニテモカメテ對手トナシ工風鍛錬スルトキハ却テ記ニ

溢ヲ得ルモノナリ

○掛声ヲ必發スベキ辯

練習中何術ニテモ必ズ掛声ヲナスベシ初心ニアリテハ

尤掛聲ノ發シ兼ルモノナリ最初ヨリ務メテ掛声ヲ發シ

習フ可シ掛声ハ我カ氣合ヲ對手ニ示シ對手ノ鋭氣ヲ挫

クモノナリ又臨機應変奇速ノ事ヲ施スニ掛声ヲ爲サ、

レハ其技遅渋シ自カラ死物トナリテ用ヲ爲サズ又掛

声ハ活気ヲ引起シ其披ヲ鋭クスルノ功用アリ

○氣當リノ大事

氣ハ生活シテ滞ルコトナク剛健ニシテ堀セサルヲ用トス

事熟スレハ自カラ氣融和シテ心常ニ温和ナリ其温和ナ

○ Always use Kakegoe

During training, no matter what technique you are doing, you should use a *Kakegoe* shout.[8] Beginners will often be loath to shout Kakegoe. However, you should train yourself to shout Kakegoe from the very beginning.

The purpose of a Kakegoe is to fracture the sharp point of your opponent's martial spirit with the force of your own martial presence. If you fail to employ Kakegoe when attempting to rapidly change your attack or defense on the fly, it will likely delay the speed with which you can apply your technique, leaving you wide open to any attack by your opponent. That is the result of not using any Kakegoe shouts. Further, using a Kakegoe will cause vitality to well up within you, meaning the techniques you employ will become sharper and more effective.

○ Striking With the Force of Your Presence

Your *Ki*[9] should infuse every action you take in your life. It is a vital source that should not be allowed to slacken. As you become more experienced, you will become more in harmony with your Ki, meaning you can maintain a calm, everyday state of mind no matter what occurs. Developing a state of mind that can remain in a relaxed, everyday state will enable you to easily channel your vitality to defeat your opponent. In a confrontation, a person with this mindset will, without a doubt, invariably emerge victorious.

---

[8] A Kakegoe is a shout unifying body and mind. Many schools have prescribed shouts, while other schools do not.

[9] Ki 気 spirit/mind/heart, is considered a fundamental element of the universe and all things arise from and are sustained by Ki. The word can also describe various psychological states. Some examples are: Expressions like:

Ki ga Omoi 気が重い"heavy-hearted"

Ki wo Tsukeru 気を付ける"be careful"

Ki ga Kiku 気が利く"to be perceptive"

Ki wo Ushinau 気を失う"lose your Ki (consciousness) to pass out"

ル沁裡ニ生育スル活溌剛健ノ氣ヲ以テ敵ニアタラハ必

ラス勝ヲ吾ニ取ルコ疑ナシ故ニ氣當リト云フヲ知ル

ベシ気當リトハ對手ト立チ合フ最初ニ於テ先ツ吾ガ常

ニ養フ處ノ剛健沈勇ノ活氣ヲ發シ對手ノ氣合ニ當リ對

手ノ気合ヲ取挫クベシ然ル片ハ對手技業縮ミ心臆シテ

勝ヲ吾ニ讓ルニ至ルベシ之レ気当リノ大事ナリ此氣當

リニ付一話アリ往時江府ニ一場ノ見為物アリ場中ノ正

面ニ犬ヶ四尺余ノ柱ヲ建テ柱上ニ一疋ノ老猿ヲ出シ其

前面ニ一條ノ誓古槍ヲ置キ箭密ヲシテ此老猿ヲ突カシ

ムルニ若シ猿ヲ突キ得ルノアレハ賞品ヲ出シ與フル

ノ則ナリ故ニ万客爭テ之ヲ突クニ老猿槍ノ發勁スル氣

ヲ察シ躰ヲ柱後ニ懸シテ空ヲ突カシム或曰種田流ノ槍

Striking your opponent with the force of your presence is known as *Ki-atari*. You should be developing your Ki during the course of your daily training. Thus, when facing off against an opponent, the first thing you should do is to forcefully project your vital fighting spirit in order to break your opponent's fighting spirit. This will not only cause your opponent's techniques to be weak and ineffective you will also instill fear into him, meaning victory will fall into your hands. That is why *Ki-atari*, striking your opponent with the force of your presence, is so important.

I would like to introduce a short story about striking your opponent with the force of your presence. Long ago, when Japan was controlled by the Shogunate, you could often find outdoor side shows.[10] One of these sideshows featured an old monkey seated atop a four-foot-tall pillar.

The owner had put a wooden practice spear out in front and challenged customers to see if they could spear the monkey. If they were able to strike it, they would win a prize. Thinking this seemed easy, great numbers of people lined up to try their luck at attacking the monkey. However, the old monkey was able to detect the moment a person was going to attack with the spear and scramble to safety behind the pillar. This meant the customer's spear thrusts struck nothing but air.

One day some students of Taneda Jubei, the famous Taneda School of Spear instructor, were passing by and decided to try their luck.[11]

---

[10] Referring to the Edo Period 1600~1868.

[11] This school was founded by Taneda Masayuki 種田 正幸 (?~?) in the Kanei Era (1624~1644.) Illustration is of a Taneda School Hooked spear.

師種田十兵衞氏ノ門人來リ試ミニ之ヲ突ク數囘老猿隱

形速ニシテ毎ニ空ヲ突カシム門生還テ之ヲ師ニ告ク種

田氏爲ニ明日門生ヲ剜テ彼ノ場ニ至リ槍ヲ取テ柱前ニ

向フ時ニ老猿苦悶一声ニ叫ンテ忽チ地ニ陷ツ主人頻ニ

シテ猿ノ命ヲ乞フ是ニ於テ種田氏微笑シテ出ツ家ニ歸

ツテ門生ニ示テ曰ク彼ノ野猿自得シテ人氣ヲ察ス故ニ

槍ノ發動スル氣ヲ察シテ早ク身ヲ柱後ニ隱クス因テ予

ハ隱ルヽ處ト共ニ彼ノ野猿ヲ突キ貫ラヌクヘキ氣

ヲ發シタレハ彼レ身ノ置處ナクシテ叫テ地ニ陷タリト云ハ

レシトゾ之則チ氣當リ妙ト云フ可キナリ

○體當リノ大事

撃劍上ニ體當リト云フアリ双方立向ヒ撃込ミタル時体

Though the students of the spear school tried several times to stab the monkey on top of the column, it always escaped to safety behind it, meaning their spears struck nothing but air. Upon their return to the Dojo, the students told their teacher about this. The next day, Taneda Jubei, along with several students, went to the sideshow. When it was his turn, Taneda, the spear master, gave a shout of such ferocious and terrible energy that it caused the monkey to drop to the ground in fright.

The owner of the show jumped forward and begged the spear master to spare his monkey. Taneda simply laughed, gathered his students and returned home. Back at the Dojo, he summoned his students around and spoke to them,

That wild monkey has developed a special ability. It can detect when people are going to react. That's why he's able to scamper to safety behind the column so quickly. However, when I struck with the force of my presence the monkey understood that, even if he escaped behind the column, I intended to pierce all the way through it and into him. When that wild monkey realized this, he screamed and dropped to the ground. This is an example of the inner secret of Ki-atari, attacking an opponent with the force of your presence.

○ The Importance of Body Checks

When doing Gekken, Japanese Fencing, there are times when you will body check your opponent. During any duel, there may be times when both you and your opponent cut at the same time, resulting in your swords blocking each other.

ト躰トヲ嘗テヽ吾ガ勇敢ヲ示シテ對手ノ活気ヲ取挫キ

自カラ驚懼ノ心ヲ生セシムルナリ之レカヲ以テ當ルニ

アラズ毎ニ養フ処ノ剛健不屈ノ體力並柔術ヲ以テ當ル

ナリ便歌ニ曰ク勝負ハ體とたいとな身のはせつヽ肉身

よりも切つ先とたせ之レ他ナシ撃合ヒ突合ヒ體當リ抔

ノ技業ヲ施ストキ吾肉身ヨリモ勇敢不屈ノ切先ノ如キ

勢ヒヲ出シ對手ノ活気ヲ取挫ク可シ

○上達者ノ心得

技術ノ上達スルニ及ヒ稍モスレバ自負剛慢ノ心出ル物

ナリ常ニ眼ヲ張リ肩ヲ怒ラセ傍人ナキ若ク為ス者ハ眞

ノ上達ニ非サルナリ武術豈腕力而巳ニ在ランヤ慎ム可

シヽヽ性時ヤ業術ノ門生漸々上達スルニ及シテ大ニ慢ス

When in such a situation you should continue the attack by slamming your body into your opponent, thereby demonstrating your martial bravery and shattering your opponent's mental resistance, giving rise to fear and doubt.

However, this is not about relying on physical power when body checking your opponent. Instead, it is combining the vital and indomitable physical strength you have been consistently developing during training with Jujutsu.

There's a folk saying that boils this all down to one point,

While in theory a battle consists of one body against another, in fact victory or defeat can be calculated not by flesh, but by how the tip of the sword moves

What this means is to use your martial force of presence to blunt your opponent's attack and break his spirit. For example, during a one-on-one sword duel, after you and your opponent have crashed into each other, you will attempt to employ a technique to take your opponent down. When doing this remember that it is your steely bravery that will decide the outcome of a duel more than your physical strength.

○ A Word of Caution to Experts

When martial arts practitioners reach expert level, there is a tendency for them to become somewhat proud and boastful. They begin glaring at people and walking around with their shoulders squared, ignoring anyone in their way. People that are full of themselves like that are never going to reach expert level as they believe mastery of martial arts is due to how strong their arms are. This is something you should be cautious about.

Long ago when Jujutsu was referred to as *Yawara*, the soft and flexible art, a certain student was extremely proud that he had become, what he considered to be, an expert. His thinking was,

I've endured brutal training for many years and have mastered the inner mysteries of this art.

However, he felt that he was unlucky since he had never had a true test of his abilities. Eventually, he decided to set out one dark night for *Nihon Zutsumi*, Japan Dyke,[12] and test his throwing skills on an unsuspecting passerby.

Soon the practitioner was delighted to see a gentleman walking down the dyke from the east. As soon as his target got into range, the practitioner stepped from the shadows and started walking towards the man as if he were going to pass the gentleman on his right side.

As the distance closed, the practitioner reached out and grabbed the gentlemen's right wrist and executed a perfect throw. However, despite being thrown through the air, the gentleman reacted by rolling and quickly standing to his feet. He then continued walking casually down the road as if nothing had happened.

The practitioner was dumbfounded at the result of his best throw and said to himself,

That is not at all what I expected...

And with that his whole mindset changed and he went home.

---

[12] Illustration on previous page of Nihon Zutsumi from *One Hundred Famous Views of Edo* 名所江戸百景 Hiroshige (1797~1858.) The dyke saw heavy traffic since it passed by the Yoshiwara Red light district. The area was destroyed in the 1927 Kanto earthquake.

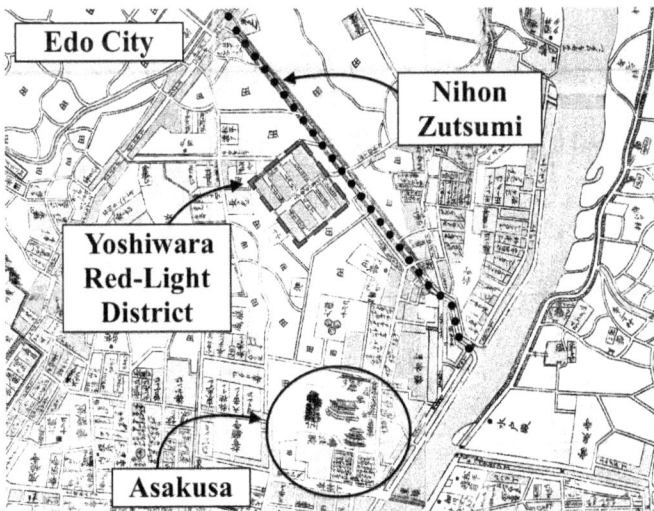

ント思ヒシガ如此白痴者ヲ殺サンハ之無益ノ殺生ナリ

ト思ヒ止リヌ彼ノ白痴者予ニ逢ヒタルハ倫數ノ尽サル

所歟若シ彼ニ均シキ無分別者ニ逢ハ彼ハ即時ニ一命ヲ

失フ可キナリ予カ門人ニハカヽル白痴ハ在間敷ケレト

モ向後ノ心得旁話シ置ナリト云ハレケリ此時夜前人ヲ

投タル柔客此ノ列ニ在リ初テ昨夜ノ士ハ吾ガ師ナルコ

ヲ知リテ惣身汗ヲ流シ黙止居タリトゾ實ニ慎ム可キハ

慢ノ一字ナリ

○附石ノ辨

技業上達シテ祐尚ホ心得ヘキ一アリ入ハ動物ナレハ譬

ヘ武術ハ上達スルモ変ニ逢ヒテハ一時動セサル可ラズ

変ニ逢ヒテ動ゼザレバ則チ死物ナリ此動スル心ヲ押鎮

The following day, at the Dojo, the Sensei called a halt to training and sat down in Seiza before inviting all his students, both novices and veterans, to gather together for a discussion. He said,

Last night I was running an errand and was walking down Japan dyke. As you know, there are a lot of violent and stupid people out in the world. One such man suddenly seized my hand as I was walking along and threw me.

As I was being thrown, I considered kicking the man to death, but then realized killing such a stupid man would be a senseless act of murder, so I held back. Further, that fool is incredibly lucky to have encountered me instead of someone else. It no doubt saved his life. If he had thoughtlessly chosen to attack a different person, he would have most assuredly immediately lost his life.

Since I believe it is not at all unlikely that my students will encounter a similarly foolish person in the future, I thought I should tell you about this episode.

It was not until the practitioner heard this talk that he realized the "gentlemen" he had thrown last night was, in fact, his Jujutsu teacher. He sat there silently, drenched in sweat. The lesson is, the thing you should be most cautious about is what is represented by the kanji *Anadoru* 慢 meaning to underestimate.

○ Placing a Stone on Your Mind

This is advice for those that have reached expert level. Humans are *Do-butsu*, creatures that move. If you are an expert in martial arts and you encounter an unusual opponent, or situation you should not be rendered unable to move. If your body freezes in place when faced with an unexpected situation, then it means you turn to a *Shi-butsu*, a dead-thing.

ムルニ又タ素ノ不動心ナリ此ノ間髪ヲ入ルヽヲ[ユルサ]ケ此

動心ヲ鎮ムルニ附拓ト去術アリ附石トハ動心ヲ引止ル

重リナリ譬ハ人アリ夏日窓前ニ坐ス迅雷已ニ起ツテ突

然我目前ニ霹ス此時ニ当ツテ何人カ驚カサラン又或ハ

市中ヲ漫歩スルニ人アリ劔ヲ抜テ我カ面ヲ突カントス

此ニ於テ何人カ心ヲ動カサ、ランヤ唯其動心ヲ押鎮ム

アリ之ヲ鎮ムレハ乃チ素ノ不動心ナリ動心ヲ鎮ムル心

ヲ仮リニ附石ト云フ附石トハ心ノ重リニ附シタル石ト

云義ナリ

○無念無想ノ辨

技業ハ心形ニツニ成レハ自在ヲ為スコト能ハス心形一致

シテ無念無相トナル無念無想トハ所謂性ヲ恣レ死ヲ忘

On the other hand, if you have a distracted mind that is unfocused and wandering, it must be calmed and brought under control. You should be in a state of *Fudoshin*, Immovable Mind, your original, unencumbered state of mind. There is a saying that goes,

You should not allow even a hair's width of space between an opponent's action and your response.

The technique for calming your mind and eliminating distraction is known as Placing a Stone on Your Mind.[13] The concept of Placing a Stone on Your Mind refers to a method of forcing your mind to stop.

Imaging that on a nice summer day, you are sitting by a window when suddenly there is a crack of lightning and a flash. In such a situation, nearly everybody would be startled. Or perhaps you are strolling around in the city when a man suddenly draws his sword and stabs straight at your face. Most people, when confronted by such a surprising situation would be unable to prevent their mind from moving, instead they would be focusing on first this, then that.

However, if you are able to calm your mind and force it down, you will have mastered *Fudoshin*, the Immovable Mind, or a calm and collected presence. By Placing a Stone on Your Mind, the weight calms you and prevents your mind from moving. This is the underlying principle.

○ Freeing Yourself From Distracting Thoughts

When applying a technique, if your attention is divided, you won't be able to execute it effectively. It is important to maintain a state of *Munen Muso*, which means Freeing Yourself From Distracting Thoughts. This involves forgetting concerns about life, death, the opponent, and even your own self.

---

[13] The word *Fuseki,* Placing a Stone, is typically use to refer to the small black and white stones used in the game Go.

レ轍ヲ志レ餓ヲ志ルレナリ斯テゾ讀ノ妙所ヲ得ルト

云フ可シ理哥ニ曰ク「るいらきをや邪とへだられてもあし今

多同し谷川の氷又「己が流れ都染をとらしそて

ぞかさり修りありけり」取浅キ教ヘニ似タレ厄着ヨ撫夫

アリ重薪ヲ頁テ谷間ノ細道ヲ歩シ瓦師ハ數丈ノ高屋ニ

ケン卑賤ノ蝶ト雖モ尚ホ如此况ヤ武術ニ於テヤ

登ヲ安ンンテ瓦ヲ敷ク之レ心形一致無念無想ニシテ自

カラ其妙所ニ至ルニ非ズンハ如何デカ此業ヲ為シ得ベ

○極意ノ辨

改ヲ修メ武ヲ講スルハ之レ修身齊家ノ原素ナリ武術ヲ

以テ勝須ヲ決シ腕カヲ磨クモノト誤認スルフ勿レ往昔

二人ノ劍師アリ或人甲師ニ問テ曰ク君等劍法ヲ練磨ス

This is what is meant by Freeing Yourself From Distracting Thoughts. There is a folk song that goes,

Rain, hail, snow and ice are all formed in different ways, however in the end they all return as river water flowing down the valley.[14]

Also,

I do not know where my art leads, however I will follow it, training to the end of my days

You can see this concept demonstrated by people who have not been formally educated. For example, a woodcutter is able to hoist a heavy bundle of sticks on his back and walk up and down hills and along narrow trails easily. A roofer can effortlessly climb dozens of feet up the side of a building and lay roof tiles accurately. These people are able to unify their mind and are demonstrating *Munen Muso*, the state of being free from distracting thoughts. Failing to reach this level of mental discipline means that your technique will remain poor. Clearly this applies to martial arts as well.

---

[14]This poem is by Ikkyu Sojun 一休宗純 (1394~1481) from *Ikkyu's Skeletons* 一休骸骨 Corresponding illustration below.

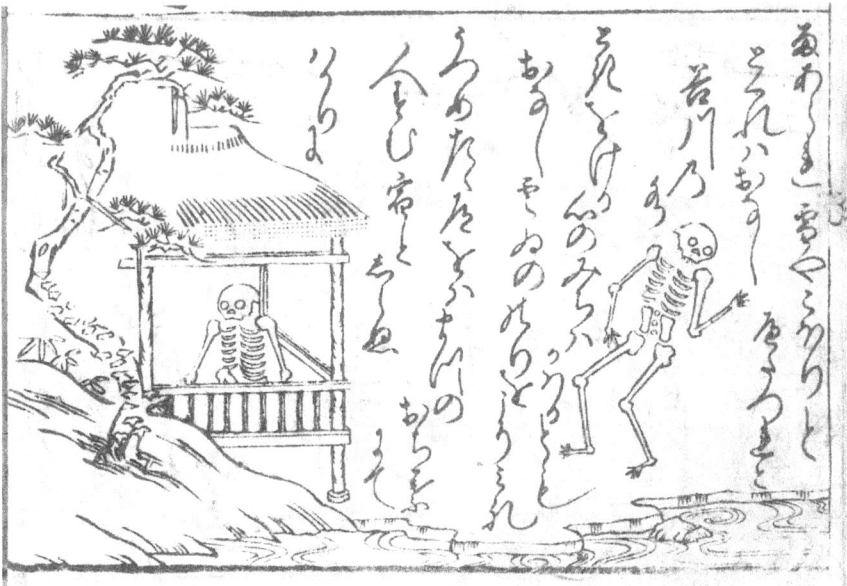

ル是ニ年アリ之レ何ノ用ニカ備フ甲師曰ク吾細夜鍛錬

スルハ人ニ勝ヘキヲ腰ストヌ乙師ニ問フ乙師曰ク吾ハ

人ニ員ケザラン事ヲ要スト人乙師ノ言ヲ嘉賞ス后チ一

人ノ劔師之レヲ傳聞シテ曰ク二師ノ説唯タ勝須ノ二事

ニ關スルノミ予ハ然ラス劔道ヲ学シテ身ヲ修メンコヲ

圖ルト宜ナルカナ此ノ言ヤ

試合ニ可打塩合ノ事ヲ弁シ

双方相對メ仕合ニ望ムハ二無尽ノ変動生スル也懸中一

待ッ中ニ掛ルヲ覚知スベシ之ヲ覚知スレバ掛待一

致ニ至リ業熟シテ理ニ至リ理究ッテ業ニ至ル時ハ業理

一体ニ至ル斯ノ如ク神龍ノ至膝ニ通暁スレバ打ニ始ナ

ク出ルニ形無所謂九天九地出入スルガ如シ斯無念無想

The purpose of book learning and martial arts training is to lay the foundation for managing yourself and your household.[15] It is important to understand that the primary objective of martial arts is not to build physical strength for the purpose of overcoming opponents in combat.

Long ago, there were two great sword masters: the Former Master and the Latter Master.

A man went up to the Former Master and asked him,

After studying the Way of the Sword for so many years, of what use is the knowledge you gained?

The former teacher replied, *I have trained night and day with great intensity in order to defeat my enemies.*

The man then asked the Latter Master the same question and he replied, *I use my training in order to not be defeated by my opponents.*

The man decided that this answer was the most compelling. Later, another sword master asked the man about his decision and the man replied,

Both teachers talked about winning and losing duels, however my goal is to learn the way of the sword while strengthening myself both physically and mentally, thus I chose the path the Latter Master described.

○ The Best Time to Strike in a Duel

When you are facing off against an opponent, there are limitless possibilities as to how each of you can attack or defend. While he is seemingly preparing to attack, he is actually on the defensive, attempting to draw you out.

---

[15] *Shushin Seika Chikoku Heitenka* 修身斉家治国平天下 is a proverb from the *Book of Rites* 礼記

Those who wish to rule the land must first cultivate their own characters, then manage their families and states, only then can they bring peace to the land.

ノ境ニ至ツテハ音モ無臭モ無鬼神モ窺フコト能ハザル

也譬バ鏡ハ物ヲ写ス故ニ心ニ警心鏡ト云ソ敵ノ心ヲ鏡

ニ掛テ見レハ打タルヽト無レド写ス心アレバ真ノ心鏡

トハ謂難シ鏡ト云モノアレバ鏡ト云一物アリ依テ写ル

ベキ鏡無シテ映ルコソ誠ノ心ノ極意ナリケレ古歌ニ曰

うつさじと思ふ心にくもらねば我も心もみかゞみにそ是名

人上手ノ位地ナリ叔敵ヲ打ベキ塩合ト云フヲ如何ナル

所ゾト云ニ常人ノ試合ニハ起ル頭ノ顕ハルヽ物ナレハ

之ヲ打ベキ塩合ト云フ之ノ敵ノ心ヨリ目ニ及処地次ニ敵

狐疑ヲ生スルヲ打ベシ次ニ居付テ動カザルヲ打ヘン次

ニ掛リ口ヲ打ベシ次ニ敵ノ心ヲセカセテ打ベシ

仕合ニ八箇條ノ法アル論

While seeming to be on the defensive he is actually readying his attack. You have to be able to determine your opponent's true intent. Being able to respond effectively no matter approach your opponent takes is *Ken-Tai Icchi*, unification of offense and defense.

By mastering martial arts techniques, you arrive at the underlying principles. Through intensive study of the underlying principles of martial arts and military strategy you arrive at techniques. The phrase *Gyo-Ri Ittai*, refers to a person who has mastered both the technique as well as the underlying theory. Like a divine dragon, you are suffused with a thorough knowledge that allows you to strike without giving the slightest sign of your intent. Your art becomes undefinable, both all-encompassing but impossible to locate.

If you reach this level of *Munen-Muso*, a state where your mind is clear and not distracted by anything, you will be immune to both sounds and smells and truly reach the level of an angry demon god.

For example, mirrors reflect things. Taking this a step further, think of your mind as being a mirror. If your opponent's intent is reflected the mirror of your mind, then, since you know what actions he will take, his strikes will all miss. However, if your intent is reflected, then everything you try to do will be instantly apparent. Thus, you should endeavor to reach a state where nothing is reflected. If you reach this level then you have attained the ultimate plateau of learning.

There is an old song that goes,

Even though I see its reflection, I don't think it is the moon. Even if it reflects, I don't think it is water. Sarazawa Pond[16]

This is the realm where true masters of the art dwell. So then, for a typical duel, the best advice is to strike the first place that occurs to you. If it is a *Shio-ai*, "Meeting of the Waves" or physical contact, then you should focus on his eyes more than his spirit. Attack him where he seems hesitant. Strike when he has stopped and isn't moving. Attack the moment he is beginning to shout. Strike when he is hurrying to recover.

---

[16] A famous pond in Nara made in 749 during the Nara period.

一ニ敵ノ粧ヲ見テ其人ヲ知ルベシ二ニ敵ト立合ノ片其

位ヲ知ルベシ三ニ自然ト業ノ移ルヲ見テ敵ノ癖ヲ知ル

可四ニ敵ノ起頭ヲ見テ我心ヲ以テ押ユベシ五ニ敵ヲ

ト間合ヲ遠クシ我ヨリ打込間ヲ近クスベシ六ニ敵ヲ釣

出スコ七ニ敵強ク守ラハ虚實ヲ以テ敵ノ心ヲ動カス

八ニ敵ノ手元強クカラカハズメ英氣ヲ避ケ虚ヲ窺ツテ

仕掛ケルコ之ヲ強敵ニハ二三ニテ勝ベシト去ヘリ

アゲコソ大切ナレ兵法ニ敵ヲ切崩ス圧芝居ヲ越ユル

勿レ追崩ストモ引揚ル堺ヲ知ルヘシト云ヘリ

心氣力一致ト云コ附テ知ノ教ノ事

心気力一致ト云コ有心気ハ理ナリカハ業ナリ此心気力

別レ〱ニナル片ハ劔道ノ用ヲ成サズ譬ヘバ火薬ノ如シ

The Eight Rules of Dueling

1. If you can look at your opponent and understand at a glance his overall presence and his disposition you will be victorious.
2. At the beginning of the duel, read your opponent's stance.
3. Watch how your opponent naturally transitions from one technique to the next, which will allow you to determine his habits.
4. Observing your opponent's face when he is relaxed will enable you to decide which way to focus your attack.
5. You should position yourself so your opponent feels you are out of his reach, while simultaneously being within easy striking distance of him.
6. It is important to draw your opponent out.
7. If your opponent remains firmly on the defensive, then use feints to force him to concentrate on you.[17]
8. When facing an opponent of great strength and ferocity, avoid attacking him directly, where he is most dangerous, and instead seek to draw him into a trap.[18] You should be able to defeat a strong opponent after a short time using this approach.

In conclusion, the important lesson in sword fighting, and all martial arts, is to break your opponent by seeing through any attempt to distract you with a strategem, and pursue your opponent until he is defeated. It is important to understand what has been outlined here.

---

[17] Kyo-Jitsu 虚実 falsehood and truth, a feint concealing your true intent.

[18] *Now a soldier's spirit is keenest in the morning; by noonday it has begun to flag; and in the evening, his mind is bent only on returning to camp. A clever general, therefore, avoids an army when its spirit is keen, but attacks it when it is sluggish and inclined to return. This is the art of studying moods. Disciplined and calm, to await the appearance of disorder and hubbub amongst the enemy–this is the art of retaining self-possession. To be near the goal while the enemy is still far from it, to wait at ease while the enemy is toiling and struggling, to be well-fed while the enemy is famished–this is the art of husbanding one's strength.*

-*The Art of War* By Sun Tzu, Chapter 8 : Maneuvering
Translated by Lionel Giles, M.A.

火薬ハ硫黄白硝灰トノ合力ヲ以テ其勢強ク天地モ震動

スルニ至ル若シ之ヲ別々ニ為ストキハ何程ノ力モ無物也

心気力一致スルコトヲ修業スベシ一致ノ力ヲ得ザレハ全

勝ハ得難カルベシ間ニ髪ヲ入レズト云フモ心気力一致

ノ謂ナリ石火ノ機ヲ云フモ心気ノ一致ヲ謂フナリ又七

知ノ教アリ武ヲ弄ブ文ニ怠ハ全体ヲ知ザル也已ニ術ニ

惇人ヲナミスルハ智ノ及ザル也寒暑ヲ恐レ業ニ怠ルハ

克ツヲ知ザル也剣長短ヲ論ルハ得失利害ヲ知ザル也相

打ノ勝敗ヲ不知ハ気ト業トノ勝ヲ知ザル也誓古スルト

ト閑暇ノ時ト別々ニ思フハ常ノ心ヲ知ザル也真剣ト誓

古ト思相ノ変ハ心ノ治ザル故也能々エ風メ徒ニ年月ヲ

賁スコ無レハ真ノ妙所ニ至ル可シ

Regarding Unifying Mind, Spirit and Strength[19]
Additional Topic: Seven Lessons on Wisdom

There is a saying that goes, *The mind, spirit and strength should be unified.* The "mind and spirit" refers to the underlying theory of a particular technique, while "strength" refers to the actual technique. You will not be able to use sword techniques when your mind, spirit and strength are not unified.

You can think of this in the same way gunpowder is made. Gunpowder is also a combination of three things: sulfur, nitrate and ash. When combined they can create a terrible explosion that shakes the very earth, however when separated each contains almost no power. Thus, it is important to train until you can unify your mind, spirit and power. Without unifying yourself, both physically and mentally, it will be impossible to reach a level where you are winning duels consistently.

The saying, *Do not allow gap that even a single hair can fit through* also refers to unifying Mind, Spirit and Strength. [20]

In addition, the phrase *As fast as a spark emerging when you strike steel to flint,* also refers to unifying of the mind and spirit.

---

[19] *Shinkiryoku no Icchi* 心気力の一致 How you should approach a duel. Evaluating your opponent both offensively and defensively.

Mind　–　Observing how your opponent moves, doesn't move and reading his intent while deciding your approach

Spirit　–　Your fighting spirit.

Power　–　Physical power as well as movement and technique.

Another phrase that means almost the same thing is *Ki-Ken-Tai no Ichi* 気剣体の一致 Unifying the Spirit Sword and Body.

[20] Mind can be thought of as being as silent as a pool of water, Spirit can be thought of as moving like the wind and Power is the energy of a wave.

Further there are also the seven lessons:

1. If you treat martial training as an amusing diversion while simultaneously neglecting to study the literary arts, you will end up not very knowledgeable.

2. Being prideful of your technique while believing others have only common ability means only that your knowledge is lacking.

3. Being afraid of doing techniques when it is too hot or too cold and neglecting to train during those times of year, only serves to demonstrate you do not know how to overcome adversity.

4. If you are debating whether it is better to train and fight with a sword with a longer blade or a shorter blade, then you fundamentally do not understand the advantages and disadvantages of each.

5. If you don't understand why a duel ended in *Ai-uchi*, where both you and your opponent scored a decisive blow in the same instant, then you do not understand the connnection between spirit, technique and achieving victory.

6. If you consider training and relaxation time to be separate entities, then you do not have a firm understanding of *Tsune no Kokoro,* maintaining the mind in a state of constant readiness.

7. The reason you feel that there is a difference between training and entering a real duel is because you have not completely conditioned your mind.

If you train diligently, and, instead of frittering away the months and years of your life develop an approach to understand what is written above, then you will eventually develop an understanding of the subtleties of martial arts.

Jujutsu

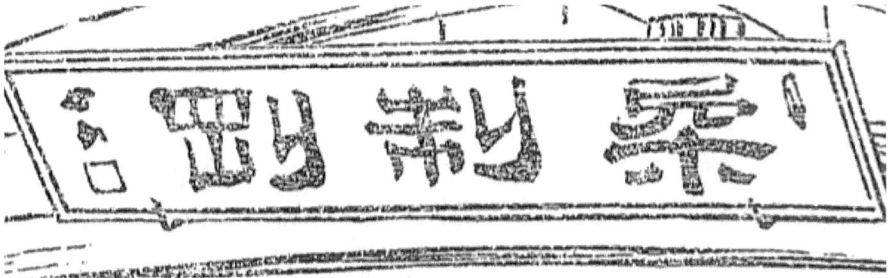

柔制剛

The soft and flexible can topple the strong and rigid.

柔術ハ古クヨリ其妙ヲ傳ヘ治亂ニ亘リ学ハ・サル可カラ

サルノ要術ナリ其術タル所謂柔能ク剛ヲ制スルノ道ニ

シテ我ニ勝レルカ者ト雖比容易ク之ヲ捕挫ギ已ヲ全フ

シテ必勝ヲ縣ツノ術ナリ夫カハ限アリ術ハ限リナシ謂

ハンヤ死活ヲ泊在ニシ其奥ヲ極ムルニ至ツテハ夭死者

メ克ク甦活セシムルアリ嗚呼該世ノ衆實ニ学バス

有ル可カラズ其流派ノ世ニ著キモノハ揚心流神道

揚流起倒流等ナリ流義ニ因リ教法大同小異アリト雖

其奥妙ニ至ツテハ一ナリ

The Origin and Significance of Jujutsu

From days of old Samurai have trained in Jujutsu. Whether in peacetime or in war, it was considered a fundamental art. What I am describing is the art where the soft and flexible conquers the strong and rigid. Even if you are facing an opponent who is more physically powerful, you can easily seize and break them. Jujutsu is an art that allows you to consistently achieve victory because, as the saying goes, *There is a limit to strength but technique has no limits.*

Jujutsu grants you the power to take life or give it. If you become enlightened to the inner mysteries of Jujutsu then you will be able to both topple your enemies and use Jujutsu resuscitation to save people who have been attacked or had an accident. Thus, I beseech the average citizen to study the art of Jujutsu.

Some of the more renowned Jujutsu schools are the Yoshin School, the Shinto School, the Tenshin Shinyo School and the Kito School. The way these schools teach Jujutsu is largely the same, varying only in the details, however the underlying theory is the same.

一柔術ハ受方ヲ甲者ト為シ捕方ヲ乙者ト為ス手術表裏

ノ鮮書ノ尽サルルハ圖画ヲ以テ辨シ圖画ノ及ハサル

ハ又書ニ讓ル讀者茅一圖ヨリ第二圖第三圖ト係連シ

テ視ルベシ仕組手解拾二組初殴居捕立合都合二拾組

投捨二拾組口傳秘決多シト虽モ大槩ネ之ヲ洩ス事ナ

シ其書画ニ尽シ難キ物ハ目録ノミヲ挙テ粗畧スル所

アリ

How to Use This Book

The person receiving the technique in Jujutsu is the Uke Kata "receiver" and the person doing the technique is the Tori Kata "engager." In this volume the receiver will be called Kosha, or Former, and the engager will be the Otsusha, or Latter.[21]

In order to make this volume easy to follow, illustrations will accompany the explanations. I have included illustrations along with the explanations. If questions remain after examining the illustrations please refer back to the text. The reader should follow the progression of the techniques from illustration one, to illustration two and then illustration three.

As some techniques have many oral transmissions and other secret teachings, they will be difficult to describe in this book. Thus, the names of these techniques will be included in the list of techniques, but not actually have an illustrated explanation.

---

[21] For clarity, the two combatants will be referred to as "you" and "your opponent."

◎鬼拳　オニコブシ

此手ハ甲方ト乙ト向ヒ合ヒ一尺余リ一図

隔テ、双方トモニ膝ニ手ヲ置

テ坐シ居リ甲ノ方ヨリ曳ノ掛

声ト共ニ乙ノ両手首ヲ圖ノ如ク

取ルヽ乙モ之ニ應シ声ヲ（ヤヽ）掛テ其持

レタル右ノ手ヲ外サントスレバ先ヅ指先ヲ延ベ甲ノ左ノ

膝ノ脇ヘ二寸押出シ直ニ其侭我ガ左ノ肩口ヘ

図ノ如ク引キ上（此時其引タル肘ガ甲ノ鼻火ト

我ガ鼻ト平三角形ヲ成ス心得ニスベレ）「ヤト」

声掛供ニ振拂フベシ續キテ

此手ハ二本続ノ手ナリ但シ右ナスハ右左ナレバ左ニ二手続

左右共ニ同シ振合ナリ

手解 Tehodoki: Freeing Your Hands 1/12
鬼拳 Onigoshi: Devil's Fist

For this technique you and your opponent are seated facing each other in Seiza, with your hands on your knees. You are about 1 Shaku, 1 foot/30 centimeters, apart.

1	2
乙Otsu　　　甲Koh You　　　Attacker 	乙Otsu　　　甲Koh You　　　Attacker

Your attacker shouts a Kakegoe of *Ei!*[22] and grabs both your wrists. This is shown in the first illustration. To free your left hand, use a Kakegoe of *Yato!* and strongly flex the tips of your fingers. Push your left hand slightly towards the side of your attacker's left knee. Then immediately yank your hand up and over your right shoulder as hard as you can. This is shown in the second illustration.

Ideally you would end up with your left elbow level with your attacker's nose. His nose, your nose and your elbow should form the three points of a triangle. Shout *Ya!* as you shake off and sweep away the hand gripping your wrist After finishing this technique you should move on to the next technique *Furi Hodoki*, Shake free, which is the second technique in the Shodan level.

Since it is best to do the technique on both sides, after doing the left hand, you should do the right hand. The right side is considered to be the second part of this technique. Both sides are done in the same manner.

---

[22] The author uses the Kanji 曳 "to pull" for this sound

一 振解 フリホドキ

此手ハ前ノ手鬼拳ト引續キ乙図

我が肩ヘ引上振拂ヒタル手ヲ直ケニ

手刀（ヤ手カシ）ヲナシ圖ノ如ク甲ノ面部

ヲ見込声ヱ七ヲカケ打込ニ甲ハ声ヲ合

セ手先ヲ矢筈ニナシ圖ノ如ク我カ額ニ当テ乙ノ

打込手首ヲ握ル乙ハ其モタレタル手ヲ又外ス二図

一ハ甲ノ右ノ膝上ニ向ケヨ—1ノ声ト共ニ図ノ如ク

打落ス。ナリ（手カヲ以献ノ撫ノ股
　　　　　ヲ裂カンズ気合ニナスゼ）

圧右共ニ同シ

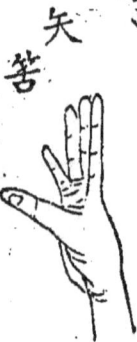

矢
筈

乙

甲

甲

乙

手解 Tehodoki: Freeing Your Hands 2/12
振解 Furi-hotoki: Shake free

This technique is a continuation of the previous technique, Devil's Fist.

1	2	*Yahazu* Nock of an arrow
甲 Koh　乙 Otsu Attacker　　You 	甲 Koh　乙 Otsu Attacker　　You 	

After using the previous technique, Devil's Fist, to yank your right hand free, strike Men, the top of the head, with that same hand. Your hand should be in a *Shuto*, or Hand Sword. This is also known as *Hira Shuto*, or Flat Hand Sword.

As the illustration shows, you aim for your attacker's head and strike with a shout of *Ei-ya!* The attacker matches your shout with one of his own and catches your wrist with his hand shaped like a *Yahazu*, the nock of an arrow. See the separate illustration of how this is done.

To free your right hand from the attacker's grip, use a cry of *Eih!* and cut your right hand downward, aiming for your attacker's right knee. This is shown in the second illustration.[23]

This technique can be done on the right and left side in the same way. When attacking with the Shuto, envision yourself striking down with enough power to snap the bone at the base of your thumb.

---

[23] Illustration two was reversed in the original document, it has been corrected here.

逆手（ギヤクデ）

此手ハ双方共ニ前ハ同シク對坐
シ膝ニ手ヲ置キ甲方ヨリ曳ト
声掛ケオカノ両手首ヲ逆ニ
圖ノ如ク唯ト握ル乙方モ
之ニ應シ〔エイヤ〕ノ声ト共ニ其持
タレタル両手ヲ外サントスルニハ先ツ
体ヲ低クシ我カ面部ヲ甲方ノ胸ノ
辺ニ当ル位ニナシヤト声掛ケカヲ
挺メテ両手ヲ左右ニ図ノ如クニ
開クナリ斯ノ如クナス時ハ容易ニ
外ルヽナリ

手解 Tehodoki: Freeing Your Hands 3/12
逆手Gyaku Te: Reverse Grip

This technique uses both hands and is similar to the previous two techniques since you and the attacker are seated facing each other with your hands on your knees.

1	2
乙Otsu　　甲Koh　　You　　Attacker	乙Otsu　　甲Koh　　You　　Attacker

As the first illustration shows, the attacker shouts *Ei!* and grabs both your hands in a reverse grip, with his thumbs down and his palms facing outward. You respond by shouting *Ei-ya!* and pushing your face towards your attacker's chest, thereby lowering your body. Your head should nearly strike your opponent in the chest. Next, with a shout of *Yato!* whip both arms out to the side as shown in the second illustration.

逆指（キビス）

此手ハ前ニ同シク双方對坐
シ膝ニ手ヲオキ甲方ヨリ
〔ヤ〕ト声掛ケ共ニ乙ノ手先
ヲ握リ指ヲ〆上ケントス
乙方ハ之ニ應シ声ヲ掛ケ
我カ肘ヲ腰脇ニ當テ拇指ヲ
屈メテ挿図ノ如ク取止メテメサセ
ヌナリ

甲

乙

手解 Tehodoki: Freeing Your Hands 4/12
逆指Gyaku Yubi: Bending the Fingers Back

This technique begins the same as the previous technique. You and the attacker are sitting facing each other with your hands on your knees.

With a shout of *Ya!* the attacker grabs your fingers, squeezes and tries to jerk them up. Respond with a shout and press your elbows into your sides, squeezing your armpits tight. Bend your left thumb towards your palm and squeeze.

This is shown in the detail illustration. This action will stop the attacker from being able to bend your fingers back.

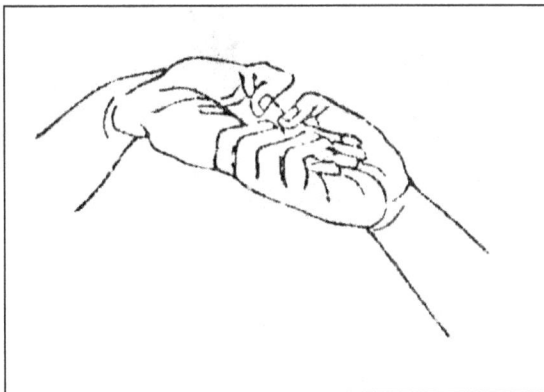

片胸捕 カタムナドリ 左右共同シ載テス 但シ此図ハ右胸捕ナリ

此手モ前ハ同シ甲ヲ右ヨリ声ヲ發ケテ 乙

方ノ胸ヲ右ノ手ニテ掴ハ乙方ハ此気ニ架

テ声ヲ發シナガラ 二図ノ如ク同ニテシ二図加甲手ノ

下ヲ我ガ右ノ手ニテ握リ左ノ足ハ

膝ヲ突ツ立ヲナシ右ノ足ヲ後ヘ引

我ガ躰ヲ半身ニ搆ヘ第二図ノ如クニナレ

左ノ手ヲ手カニナレヲ甲ノ掴タル手首

ニ押当（エイヤ）ト声ト共ニ上下ヘ引分ニナシ手カ

ニテ押落シ右ニ握タ 我ガ肩マヘ

ルナリ

引上

図

二

引上

手解 Tehodoki: Freeing Your Hands 5/12
片胸捕 Kata Mune Dori: One Handed Chest Grab

This technique is done the same on both the left and right sides, however the illustrations only show how it is done on the right side. *Defense Against One Handed Chest Grab* starts out the same as the previous one.

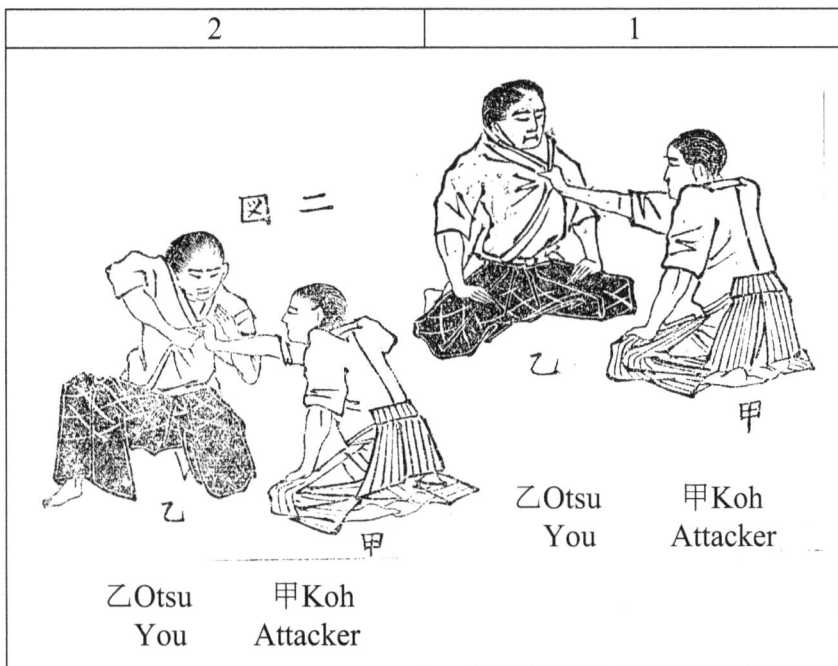

2	1

図二

乙Otsu
You

甲Koh
Attacker

乙Otsu
You

甲Koh
Attacker

The attacker shouts and grabs your collar with his right hand. You use this attack to gather your strength and shout in response. Then do as shown in the illustrations.

With your right hand, grab your shirt just below where your attacker grabbed you. Focus all your weight on your left knee and pull your right leg back. This will cause your body to rotate clockwise. You are now perpendicular in relation to the attacker.

Form your left hand into a Shuto, Hand Sword, and with a shout of *Eiya!* strike your attacker's right wrist. You are simultaneously striking the hand holding your collar from above with your left hand and pulling up with your right hand from below. Your right hand should pull diagonally up towards your right shoulder.

両胸捕 リャウムナトリ此図ハ右両胸捕ノ図ナリ

○此手モ前ハ同ヤウナリ甲ノ方ヨリ声ヲ

（ヤレ）掛ヶ乙ノ胸襟ヲ双手ニテ取ル乙ハ掛

声ニ答ヘ我ガ腰ヲ少シク延ベ所謂
中ゴシ

（甲ノ右ノ手上ニ有ケハ右ノ両胸捕ト号シテ左

ノ手上ニ有ケハ左ノ両胸捕ト号ス）図ノ如ク我ガ

左ノ手ト右ノ膝ヲ立左リヲ膝突ツマ立ルナリ甲方ノ

両手ノ上ニ押当（ヤレ）ト掛声ト共ニ少シ腰ヲ下ル途端ニ甲ノ両手

ノスキガアク故図ノ如ク我ガ左ノ手先ヲ差入捕画ノ如ク我ガ右ノ

手ニテ握リ（ヤレ）声ト共ニ下ヨリコジ上ルナリ（手鞹ノ左ノ
　　　　　　　　　　　　　　　解ルナリ）其途

端ニ左右ノ足ヲ立替右ノ足膝突左リ足ヲ立膝ヲナレテ残リ

タル甲ノ右ニテ、捫タル手ヲ前ノ片胸捕ノ如ク二打拂フベレ左右共同ジ

乙

甲

手解 Tehodoki: Freeing Your Hands 6/12
両胸捕 Ryomune Dori:
Defense Against a Two-Handed Chest Grab

This illustration shows the technique *Defense Against a Two-Handed Chest Grab* on the right. This technique begins from the same position as the previous one.

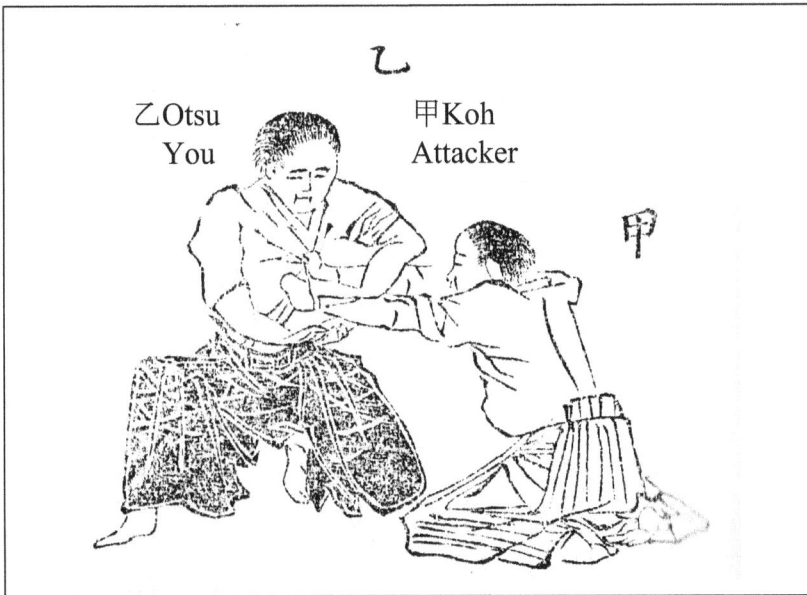

The attacker shouts *Yaa!* and grabs your left and right collar at chest level with both hands. Respond with your own shout and raise your hips up off the ground. The action of raising your hips up off the ground like is known as *Naka-goshi*, raising yourself half-way up.

If your attacker's right hand is on top, then the technique is called Defense Against a Right Two-Handed Chest Grab, if his left hand is on top then it is called Defense Against a Left Two-Handed Chest Grab.

Next, put your weight on your left knee and stand up on your right foot while raising your left hand. Push down on the top of the attacker's arms with your left hand. Shout *Yaa!* and, at the same time, drop your hips slightly.

The moment you do this a gap will open up between the attacker's arms. As the illustration shows, make use of this chance and slip your arm in between your opponent's arms and clasp your hands together.

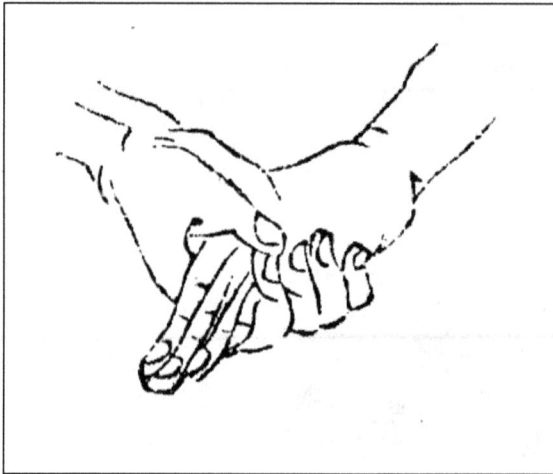

Join your left hand to your right as shown in the detail illustration. With a shout of *Yato!* bring your hips up. This will cause the attacker's left hand to release. From there immediately switch your feet by dropping your right knee to the ground and standing up on your left foot. Knock off the attacker's remaining hand by striking with a Shuto in the same way as described in *Kata Mune Dori*. The left and right sides are done the same way.

Translator's Note:
Alternate Illustration from the 1893 Edition

Positioning for *Defense Against a Two-Handed Chest Grab.*

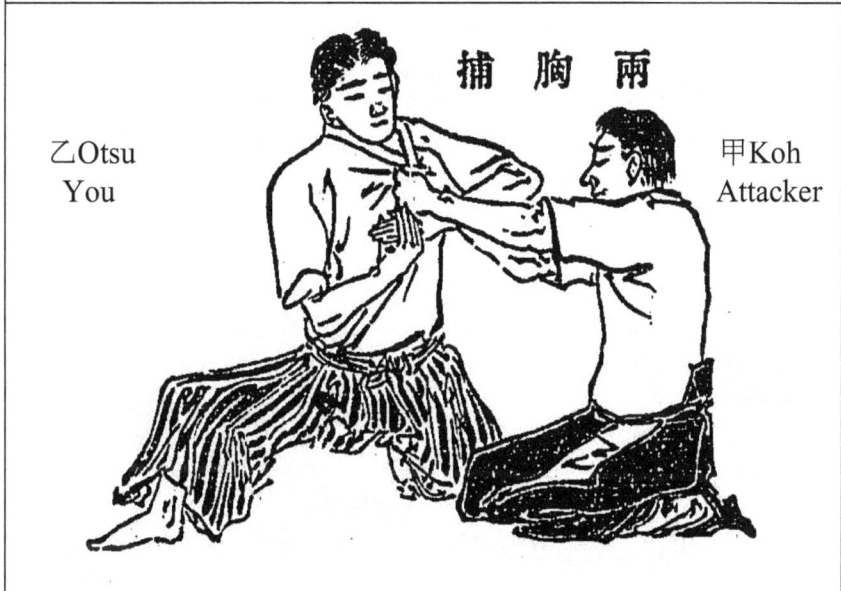

兩 胸 捕

乙Otsu
You

甲Koh
Attacker

○小手返 コテガヘシ

○此手ハ双方膝ニ手ヲ置キ對坐シ（尺余ヲ隔ツ
ベシ）甲ノ方ヨリ「ヤ」ト乙「エイヤ」ト共ニ声ヲ交ヘ
圖ノ如ク甲ノ左ノ手先ヲ右ノ手ニテ逆ニ取
リ左手ヲ持添ヘ左足ヲ甲ノ左膝ノ横ニ立テ
右膝ヲ突圖ノ如ク上ヘ持上ケ其侭声ヲ掛ケ右ヘ倒シ
甲ノ左脇下ニ左リ膝ヲ突キ右ノ膝ヲ立テ
我カ体ヲ半身ニナシ圖ノ如ク彊敵ノ手
先ヲ押サヘ我カ右ノ足ノ爪先ヲ見込ヨリ
手ノ押ヘ方ハ圖ノ如ク手背ノ紅指ト中指ノ付根ヲ
押付ルナリ甲ノ方ハ倒レルトタシニ左リナラハ左足ニテ乙ノ面部ヲ
ケル勢ヲナスベシ　左右同シ

手解 Tehodoki: Freeing Your Hands 7/12
小手返 Kote Gaeshi: Bending Back the Hand

This technique begins with you and the attacker seated facing each other in Seiza with your hands on their knees. You should be about 1 Shaku, 1 foot/30 centimeters, apart.

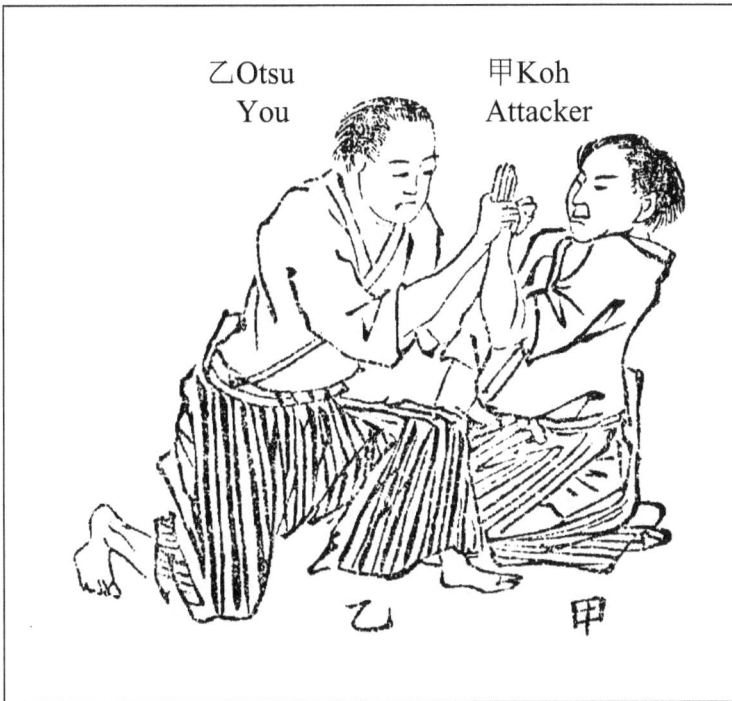

As soon as the attacker shouts *Yato!* you should respond with your own shout of *Eiya!* Then reach out with your right hand and seize the end of the attacker's left hand then twist his wrist clockwise. This action is called *Gyaku ni Tori*, Taking a Reverse.

Next, join your left hand beside your right hand and plant your left foot beside the attacker's left knee being sure to keep your weight on your right knee. This is shown in the illustration. After bringing the attacker's hand up, shout *Eiya!* and topple your opponent to your right. Finally, plant your knee below the attacker's left armpit.

乙 Otsu
You

甲 Koh
Attacker

As you rotate your body clockwise, drop your left knee down and stand up on your right foot. As the illustration shows, force the tips of the attacker's fingers into the Tatami mat. Turn your head so you are looking the same direction as the toes of your right foot, as if you are staring at them.

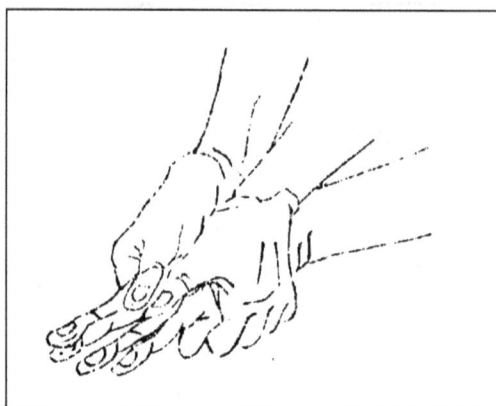

The detail illustration shows how to take a Gyaku, joint lock. It is important to note that you are using your thumbs to push into the base of the middle fingers and ring finger. The moment the attacker is toppled over, he will attempt a strong kick aimed at your face. If he has toppled on his left side, then he will kick with his left foot. The left and right sides are done the same way.

Translator's Note:

The *Kote Gaeshi* from the 1893 edition has some interesting differences. In that edition you keep your left foot and right knee on the ground, the opposite of the 1887 version. You are also looking straight down as opposed to looking to the right at your toes.

Illustrations 1 & 2 from the 1893 edition

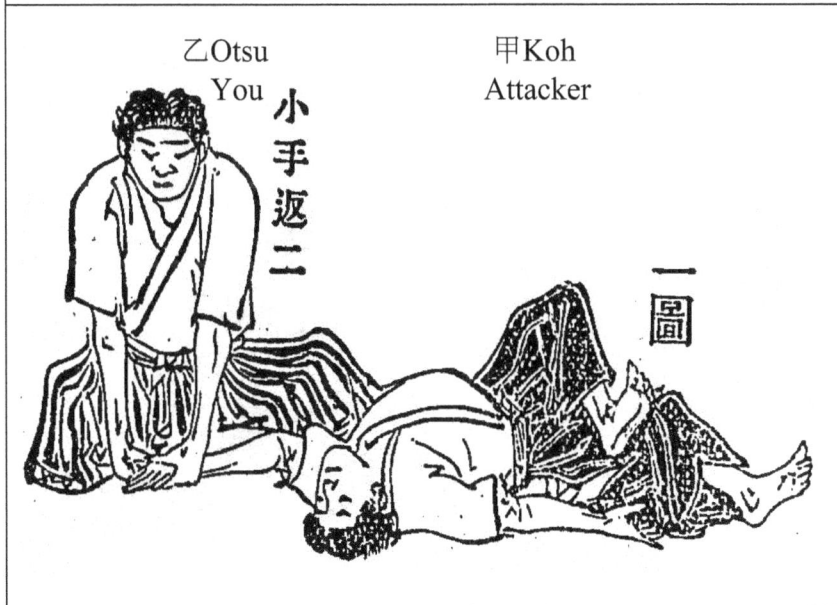

乙Otsu
You

甲Koh
Attacker

乙Otsu
You

甲Koh
Attacker

小手返二

一圖

両手返　リャウテカヘシ

〇此手モ前ハ仝シク双方膝ニ手ヲ置キテ

對坐シ甲ノ方ヨリ「アイヤ」ト声掛ケ乙ノ

右ノ手首ヲ両手ニテ圖ノ如ク唯握ル

乙ノ方ハ「ヤ」ト答ヘテ其持レタル手ノ

指先ヲ延ベテ一寸許甲ノ左リノ膝ノ

向ヘ突キ出シ直ニ我カ左リ肩ロノトコロマデ引上

ケ持返シ手ノ掌ヲ逆ニナシテ又甲ノ元ノ

膝ノ脇横ヘ「エイ」ト声掛ニ押込スナリ

左右共ニ同シ理ナリ

乙　　　　　甲

90

手解 Tehodoki: Freeing Your Hands 8/12
両手返Ryote Kaeshi: Two-Handed Reverse

This technique begins the same way as the previous technique, with both you and the attacker seated in Seiza, facing each other with your hands on your knees.

With a shout of *Eiya!* the attacker grabs your right wrist with both hands. This is shown in the illustration.

In response, you should shout *Ya-a!* and put all your power in the fingertips of the hand the attacker has seized. Thrust your right hand slightly forward, towards the attacker's left knee, then immediately yank your hand up over your left shoulder. Reverse your palm, and then, with a shout of *Eiya!* whip your arm down towards the attacker's knees.

The left and right sides follow the same principle.

Ryote Dori (Ryote Kaeshi) from the 1893 edition

捕 手 両

甲 Koh
Attacker

乙 Otsu
You

Ryote Dori (Ryote Kaeshi) from the 1926 edition

一、両手捕

Translator's Note:

As shown on the previous pages the 1893 and the 1926 editions call this technique *Ryote Dori*, Two Handed Capture. However, judging by the illustration and the fact that it is listed after *Kote Gaeshi* and before *Kidori* (the following technique) it appears to be the same technique with a different name.

The description from the 1893 edition also has some noticeable differences:

The attacker grabs your right wrist with both hands, grabbing first with the right and then the left. In response the attacker shifts his right knee back and pulls. You anticipate this action and suddenly stab your right hand forward then immediately yank it upward so your hand is above the attacker's right wrist. Ensure you stab your fingers almost to the attacker's waist.

氣捕 キドリ

○此手ハ前ハ同双方ニ尺余ヲ隔テ

向合甲方ヨリ〔ヤ〕掛嘉右ノ手ニテ一

乙方ノ面部ヘ打付ル勢ヲナス乙方モ図

其氣ニ架シテ第一図ノ如ク我カ右ノ

手ヲ差延シ甲ノ胸ニ我手ノ皆ヲ押シ

当左手ヲ甲ノ呑ク膝ノ下ニ指先ヲ差入久又我カ右

足ヲ甲ノ膝元ニ逆踏出〔ヲイ〕ル罘ト共ニ向ヘ押倒スリ

甲方ハ起〔アガ〕ラント勢ヲナスナリ乙方ハ

直ニ我カ躰ヲ右後ニ開キ第二図ノ如ク右ノ膝ヲ

立左リノ膝ヲ突キ爪先ヲ立我カ両手ニテ罘ヲ囲ヒ

十分ノ身搆ヘヲナスナリ〔恩又起立リテ掛ランフ

図 一 甲

乙

図 二 甲

乙

94

手解 Tehodoki: Freeing Your Hands 9/12
気捕 Kidori: Seizing the Chance

This technique begins the same way as the previous technique. Both you and the attacker are seated facing each other in Seiza approximately 1 Shaku, 1 foot/30 centimeters, apart.

乙Otsu　　　　甲Koh　You
Attacker

With a shout of *Yaa!* the attacker shows his intention to punch you in the face with his right hand. Realizing this, you do as shown in the illustration. Extend your right arm out and push the attacker in the chest with the back of your hand and, at the same time, slip the fingers of your left hand under his right knee.

Illustration 2

乙 Otsu　　　　　甲 Koh　　You
　　　　　　　　　　Attacker

With a shout of *Eiya!* flip the attacker backwards. If the attacker seems like he is going to recover and attack, then you should rotate your body to the right and away so you are at an angle to the attacker.

As the second illustration shows, your right knee is upright while your left knee is planted on the ground with the toes of that foot on the ground and your heel off the ground. Both hands are held low surrounding your testicles and you are in a stable and ready stance. You should be vigilant against any attempt by the attacker.

The final image of both the 1893 and 1926 editions show the arms forcefully extended.

1893 edition

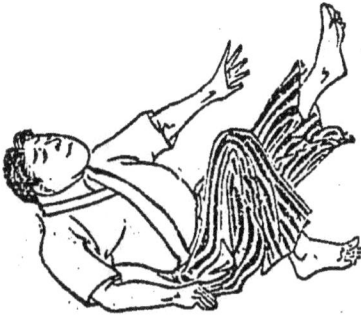 

1926 edition

天倒（テントウ）

○此手モ前同シ双方向ヒ合膝ニ
手ヲ置ニ尺余ヲ隔坐シ居甲方ヨリ
従ト掛ル第ト共ニ一図ノ如ク左リ足ヲ乙ノ
膝ノ元迄踏出シ乙テ両手ニテ抱込ム乙方ハ
（ヤ）答ヘ両手ニテ我ガ肇ヲ囲ヒ乍腰ヲ少シク
延上リ右ノ足ヲ右ヘ開キ立膝ヲナシ左足膝ヲ
突込先ヲ立テ我ガ両肘ヲ張ル此時甲ノ抱込タル
双手ノスクガ故ニ左ノ手ニテ甲ノ衣類ノ紋所ヲ掴
ミ右ノ手ヲ我ガ肩ヘ引發出シ甲ノ天倒ヘ拳ニテ
（猶グヒ）押當エイヤト言ッテ打落スナリ即チ二図
如ク拳ヲ強ヨク下ヘ押付ルナリ

一図

乙

甲

二図

乙

甲

手解 Tehodoki: Freeing Your Hands 10/12
天倒 Tento: Top of the Head

This technique begins the same way as the previous one. You and the attacker are seated in Seiza across from each other, approximately 2 Shaku, 2 feet/60 centimeters apart.

Illustration 1

With a shout of *Ei!* the attacker slides his left knee towards your knees and wraps his arms around you from the front. This is shown in the first illustration.

Illustration 2

You respond to this with a shout of *Yaa!* and drop both hands low, surrounding your testicles as you raise your hips up slightly and step out to the right with your right foot.

Keep your left knee planted on the ground but stand up on your toes, with your heel off the ground. Put power in both elbows and push out. This will cause some space to form between you and the attacker.

Next, slip your left hand under the attacker's right arm and around his back. Grip the *Mon-dokoro*, family crest, on the back of his shirt with your left hand. Raise your right shoulder up and free it from the Attacker's left arm, then plant your right fist on the *Kyusho Tento*, the top of his head.

This spot is also known colloquially as *Gunkotsu*, the soldier's bone. With a shout of *Eiya!* knock him down. As is shown in illustration two, use your fist to strongly push the attacker down.

Translator's Note: The illustrations on the following page combine the illustrations from the 1887 edition and the 1893 edition to give more complete picture of the technique.

Illustration 1 1887

Illustration 2 1893

Illustration 3 1887

扱捕 モギトリ

三図

○此手モ前ハ同クニ尺余ヲ隔テ
双方向合ニ坐甲方ハ小太刀ヲ帯（オビ）
テ構ヘ甲ヨリ（チヱ）ト掛声共ニ乙方
ノ面部ヲ見掛小太刀ヲ以テ切込ヲ
乙ハ（ヤト）答第一図ノ如ク甲ヨリ打込ム
小太刀ノ手首ヲ左ノ手ニテ掴ミ直ニ右ノ手
ヲ以テ三図ノ如ク右足ヲ甲膝元マテ進メ
左足ハ膝ヲ突（爪先ヲ立セ）直ニ（ヤト）声ト共ニ我ガ体ヲ右後ヘ
取名手ヲ寸上方ヘ差シ（此ハ甲ノ手ノ痛ス多シ）直ニ我ガ右ノ爪先迫甲
一文字ニ開キ右足ヲ我ガ右方ヘ開キ立膝ヲナス途端ニ両手
手ニ至迄ニシテ畳ニ押付擽捕（ルナリ）此時甲ハ與畳ヲ打テ敗ヲ示ス

一図
甲
乙

二図
甲
乙

手解 Tehodoki: Freeing Your Hands 11/12
扱捕 Mogi Dori: Plucking Away

This technique begins the same way as the previous technique with both combatants seated in Seiza facing each other approximately 2 Shaku, 2 feet/ 60 centimeters apart.

Illustration 1

乙Otsu
You

甲Koh
Attacker

The attacker draws a Kodachi short sword from his belt and prepares to attack. He shouts *Eiya!* indicating he is going to cut down on the top of your head. As he begins his cut, respond with a shout of *Ya!* and, in one motion, grab the attacker's right wrist with your left hand. Immediately grab the fingers holding the handle with your right hand.

Step forward and plant your right foot in front of the attacker's knees while keeping your left knee on the ground. The toes of your left foot should be on the ground with your heel off the ground. This is shown in the first illustration.

The detail illustration shows how to grip with the right hand. Dig your fingers under the fingers of the attacker holding the handle of his sword. Your thumb should be pushing at where the index finger and middle finger meet the back of the hand.

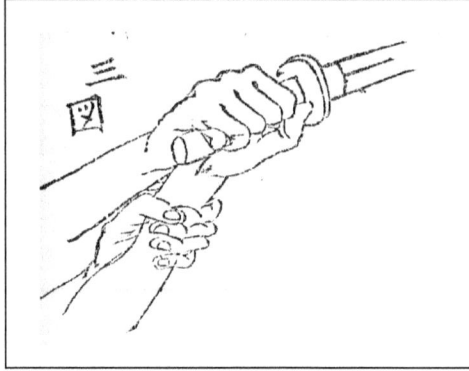

Immediately shout *Yato!* while dropping back with your right foot, rotating your body clockwise so you end up completely perpendicular to your opponent. You should be making an Ichi-monji, a straight line like the Kanji 一 meaning the number "one."

Next, as you begin moving your right foot out to the side, raise your arms up a little. This will make the next step less painful for the attacker. While moving your right foot, drag his arm towards the toes of your right foot. Pull the attacker's arm down flat to the Tatami so he releases his sword. This is *Mogi Dori*, Plucking Away. This is shown in the second illustration. The Attacker should slap the Tatami to signify his loss.

Illustration 2

Step 1 from the 1893 Edition

Step 2 from the 1887 Edition

Step 3 from the 1926 Edition

打手 ウチテ

○此手モ前同シ甲方ヨリ（エイ）ト甲

声掛ケ右ノ拳ヲ以テ乙方ノ眉間ヲ

打ツ乙方モ（ヤ）ト声ヲ答（左）リ

手ヲ三第図ノ如ク文字ヲ受止

直手首ヲ捕リ右手ヲ甲ノ

右ノ肩ニ押當テ捕リタレ

蹴ノ手先ヲ我カ左リノ腰ニ引付ケ

左右ニ引分（エイ）ト声ヲ掛ケ我ガ右ノ膝ヲ

甲ノ右膝ヲ横ニ立左リノ膝ヲ突キ

（此先ヲ立ル）第二図ノ如クナシ右ヘ廻シ礮ニ

倒シ直ニ我カ右ノ膝ヲ甲ノ右ノ腋ノ下ノ所ヘ※

一図

※突左膝ヲ立甲ノ右ノ耳下

ニ我カ右ノ手ノ大指ヲ押當

ツヨク押シ（ヤ）ト声掛ケ

左右ニ引分ルヽ

此時甲畳ヲ打テ頂

示ス

左右共ニ同シ理ナリ

二図

106

手解 Tehodoki: Freeing Your Hands 12/12
打手 Uchi Te: Striking Hand

This technique begins the same as the previous one.

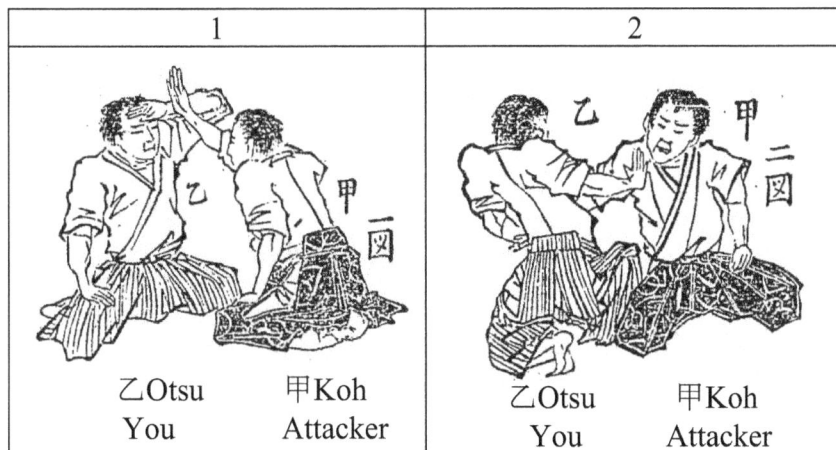

1	2
乙Otsu　甲Koh You　Attacker	乙Otsu　甲Koh You　Attacker

The attacker shouts *Eiya!* and punches to Miken, the spot between your eyebrows. You respond with a shout of *Yato!* and do an Ichimonji block with your left arm as shown in the first illustration. When blocking your arm forms a straight line like the Kanji Ichi 一 meaning the number one. Next, seize your attacker's right wrist with your left hand.

With your right hand, slap down hard on the attacker's right shoulder and grab. Next, pull the attacker's right hand down to your left hip. With a shout of *Eiya!* plant your right foot beside the attacker's right knee, push with your right hand and pull with your left. Your left knee is planted on the ground with the toes of your left foot on the ground and your heel off the ground.

As the second illustration shows, rotate counterclockwise to topple the opponent onto his right side. As you do this drop your right knee to the ground beside the attacker's right armpit and stand on your left foot. Use your right thumb to press firmly just below the ear of the attacker. Push with your right hand and pull with your left hand as you shout *Yato!*

The attacker should slap the Tatami to indicate he has been defeated. The left and right sides of this technique are done the same.

真之位 Shin no Kurai:

初段真之位 シンノクラヰ

○此手ハ双方共一間余ヲ隔
テ正面ニ向ヒ双方二図ノ如ク身構ヲ
ナシ甲方ハ両手ヲ膝ニ置乙顔ヲ睨込ミ（セ）
ト声ヲ掛ル乙方ハ双ニ應シヤート信ガラ両手ニテ

甲 一図

二ジリ〱トメ込ヨリ甲方ハ右ノ拳
ヲ以テ乙ノ面部ヲ打ントス
乙方顔ヲ右ヘ向ヶ我
カ爪先ヲ見ヨリ
都テ第二図
ノ如シ
甲

澤丸ヲ回ヒ左ノ膝ヲ立甲ノ面ヲ見込ミ淇處ニ両足ヲ揃ヘ突立
左足ヨリニ又千鳥ガヶニ甲右膝元ノ五六寸許前ヘ進ム
此三時甲方ヲ両手ヲ横ヶ抱付勢フ以ス故ニ六右ヲ指ヲ
剥臂ヲ甲眼先入ルヲ以テ甲ノ顔
ヲ左ニ向故直其手ニテ甲ノ肩ヲ

二図

乙

捕左足ヲ後ニ踏込甲後立左ノ手ニテ甲右ノ
肩ヨリ上襟ヲ掴ミ右ノ手ニテ左ノ襟肩ノ衣ヲ掴ミヤート言テ後ヘ突下リ淇時ニ右膝ヲ立テ

初段 Shodan Idori: First Level Techniques Seated Techniques
1/10 真之位 Shin no Kurai: True Stance

This technique begins with you and the attacker seated in Seiza facing each other 1 Kan, or 6 feet/1.8 meters, apart.

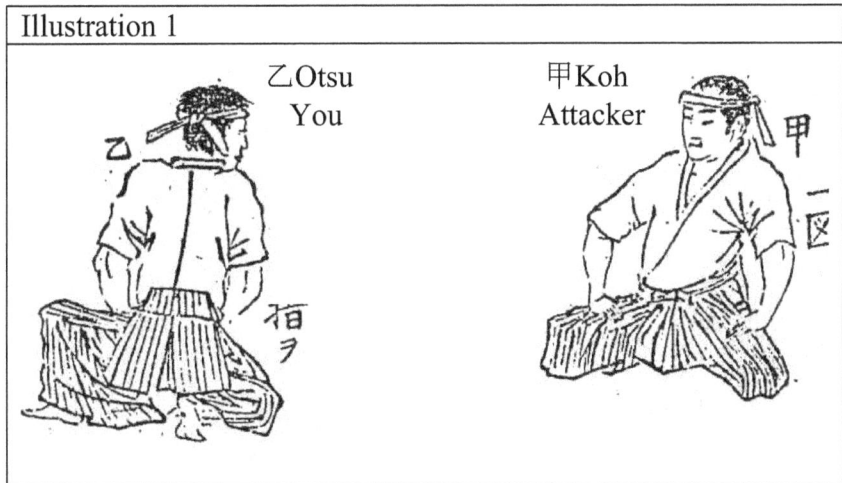

The attacker, with his hands on his knees glares at you and shouts of *Eiya!* You respond by glaring at him and shouting *Yaa!* as you press your right knee into the ground and step out to the left with your left foot, being careful to keep your hands encircling your groin. This is shown in the first illustration.

From there, bring your feet together and stand up. Starting with your left foot take two rapid steps towards your opponent. These steps should be quick like the scampering step of the Chidori, Japanese plover bird, with one foot crossing over the other. Your right foot should stop 5 or 6 Sun, 5~6 inches/15-18 centimeters, from your opponent's right knee.

In response to this, the attacker spreads his arms wide open as if trying to wrap you up. You strike by joining the fingers of your right hand together and stabbing towards his eyes. This will obscure his vision and distract him by causing him to look away to the left. Use this distraction as a chance to grab your opponent's upper arm and step behind him with your left foot.

Illustration 2

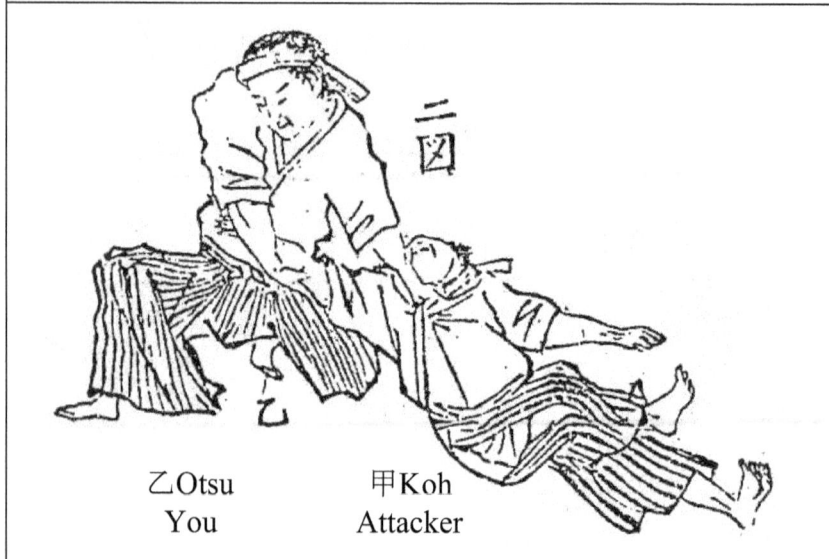

乙 Otsu   甲 Koh
You   Attacker

Seize opponent's left upper collar with your left hand and with your right hand grab his collar, shoulder or some part of his sleeve.

With a shout of *Yaa!* drop back one step with your left leg and plant your left knee on the ground. Keeping your right knee upright, use steady pressure to topple your opponent.

After being thrown, your opponent will ball up his left fist and try to punch you in the face. To protect yourself against this attack, turn your face to the right and fix your gaze on the toes of your right foot. This is shown in the second illustration.

Differences in the 1926 Edition

In the first illustration (top) you keep your left foot close beside your right foot .

In the second illustration you are holding the Attacker's right wrist.

位之眞、一

眞ノ位二圖

添　捕ソヘドリ

○此手ハ双方居並ヒ坐シ初ハ甲　一図

双方共ニ膝ニ手ヲ置テ顔ヲ見合

甲方ヨリ(エイセ)發声乙方之ニ

應シヤート言テ直ニ甲ノ右横ニ体ヲ

躬メ右手ニテ甲ノ下ヱリヲ取リ其時右足ハ膝ヲ突キ

左足ハ甲ノ猜後ニ立左リ半ヲ差延ベ右手ニ取タル下ヱリ

ヲ左ニ持替右手ハ甲ノ右ノ手首ヲ握リ一図ノ如クナシ

我ガ肘ヲ把力脇腹ニ當テ(エイヤ)ト言テ向ヘ横ニ倒レテ

敵ヲ投ケ左右手ヲ離サズ直ニ首ヲ抜キ起直リ

第二図ノ如クナシ右膝ヲ立テ両手ニ持タル襟ト

手首ヲ左右ノ腰ニ引付メ込ナリ※

※甲方ハ乙ノ躰或ハ畳ヲ打ッテ敗ヲ示ス

初段 Shodan: First Level Techniques
2/10 添捕 Soe-dori: Alongside Seizure

This technique begins with both combatants seated alongside each other in Seiza, with your hands on your knees.

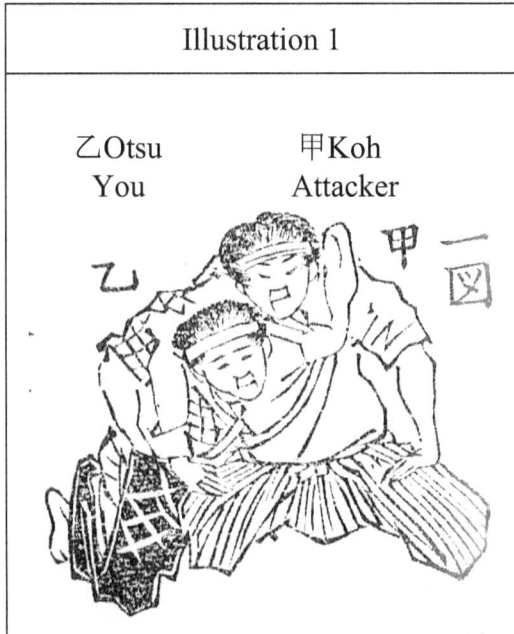

Illustration 1

乙Otsu 甲Koh
You Attacker

You and the attacker turn to look at each other and the attacker shouts *Eiya!* In response, you shout *Yaato!* and immediately shift your body towards his right knee. You then grab the attacker's lower lapel with your right hand. As you do this press your right knee into the ground and step behind him with your left foot.

As shown in the first illustration, the attacker responds by wrapping his right arm around your neck, reaching for your right hand on his collar. Release his collar with your right hand, reach over his shoulder with your left hand and grab the same spot on his collar. Seize your opponent's right wrist with your right hand.

Next, with a shout of *Eiya!* pull your elbows in close your sides, thereby toppling him beside you. Do not release your hands while throwing, however you should immediately free your head and end up in the position shown in the second illustration.

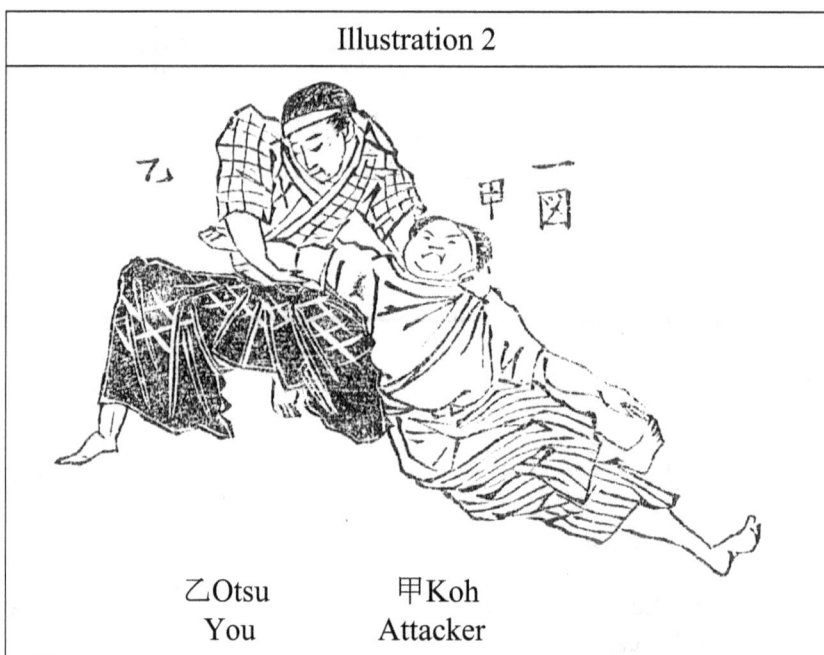

Illustration 2

乙Otsu   甲Koh
You   Attacker

The technique ends with your right leg upright and your left hand griping his collar while your right hand pulls your opponent's wrist to your hip. Your opponent should strike either the Tatami mat or his body to indicate he is defeated.

Note: *Alongside Seizure* with additional images from the 1926 edition.

1	2

3

4

御前捕 ゴゼンドリ

○此手モ双方訓坐シ膝ニ

両手ヲ置キ甲方ヨリ（ヱイセ）ト

発声シ双方共気ヲ計リ頗ヲ見合セ

乙方モ應シヤイト声ヲ掛ヶ右ノ膝ヲ

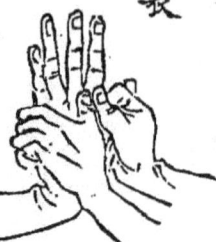

表

取方

逆手ノ
方

逆手ノ裏

俟甲ノ両手ヲ押

少シ進ミ坐シ死

乙

甲

乙

左ノ足ヨリ先ニ
立チ中腰ニテ左ヘ

左ノ我左ノ手
右ヲ押ル足

廻シ押ヘタル左ノ手ノ掌ヲ迎ニ我

左ヰヲ三ニテ取小持上右ヲ持添テ一

二図

一図

甲

左足ノ流先ニテ甲ノ刷肩ヲ

ケル（其形迄）ノ如ク立チ

我左体ヲ文字ニ横直シ

左ノ脇（左足ヲ三尺許踏

込ミ膝ヲ立上図ノ如ク甲

掌背ヲ持ヶ

我臍辺ヘ

引

付

リテ

ケル（其形迄）ノ如ク立チ

ムリ

左ヲ我肩

ト込

リ右ニ

甲

初段 Shodan: First Level Techniques
3/10 御前捕 Gozen Dori: Before A Royal Person Technique

This technique also begins with both combatants seated beside each other in Seiza, with your hands on your knees.

Illustration 1

You and your opponent stare at each other, each seeking to judge the other's intent The attacker locks eyes with you and shouts *Eiya!*. You respond with a shout of *Yaa!* and move your right knee forward so it is approximately 2 feet/20 centimeters in front of the attacker.

Then, seize your opponent's hands. (Your right hand suppresses his right hand and your left hand suppresses his left hand.)[24] Bring your left leg up and rise into *Nakagoshi*, half-way standing. As you step to the left rotate his left palm so it is facing away from you while lifting it up. Next, bring your right hand up to join your left hand as shown in Detail Illustration.

---

[24] Information in brackets is by the author.

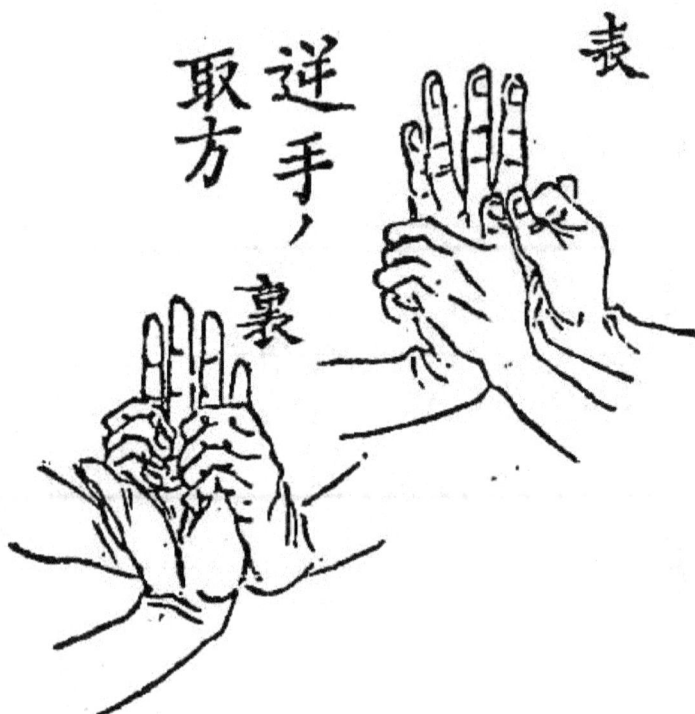

Detail Illustration:
Gyaku Te no Torikata
How to take the wrist lock.

Right: How your thumbs should be positioned on the back of your opponent's hand.
Left: View from the palm side.

After locking his wrist, kick your opponent in the ribs with the toes of your left foot. When practicing, this kick should not actually strike your training partner. It should just show *Katachi*, the movement. This is all shown in the first illustration.

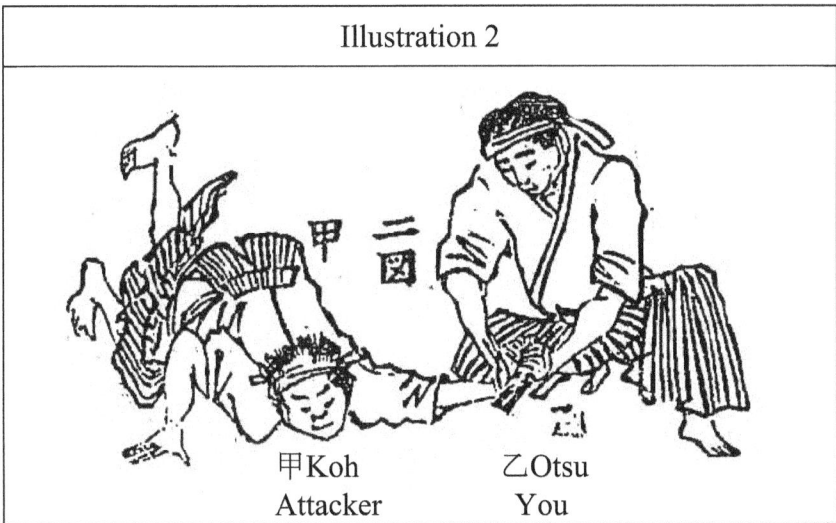

Illustration 2

甲 Koh
Attacker

乙 Otsu
You

Finally, step sideways to the left about 3 Shaku, 3 feet/90 centimeters, with your left foot and drop down onto your right knee. Be sure to maintain your hold on your opponent's left hand and pull it towards your navel as shown in the second illustration. Apply steady pressure to his wrist. The left and right sides of this technique are done the same way.

1926 Edition Illustration 1

御前捕
一

1926 Edition Illustration 2

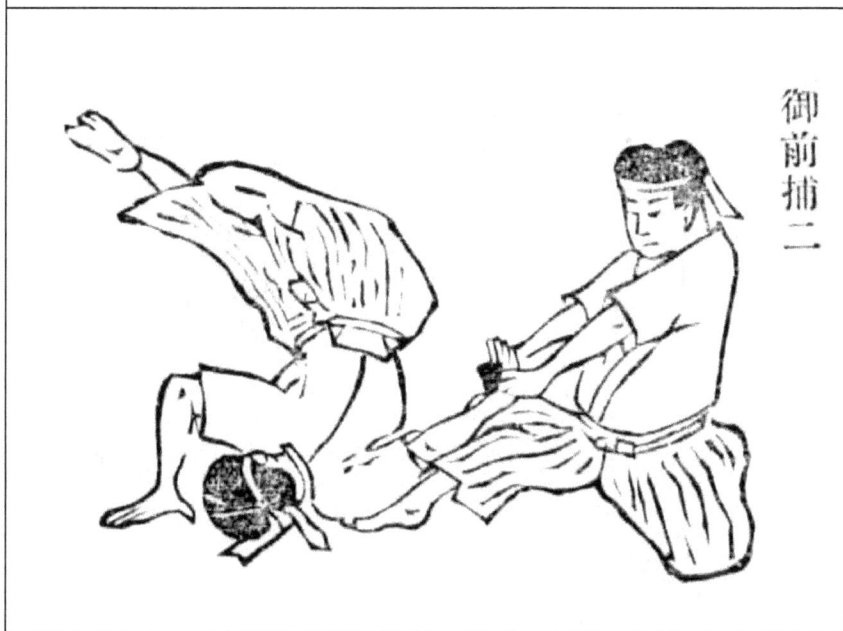

御前捕二

○袖　車 ソデグルマ

○此手ハ眞位ノ構同シク對坐シ

甲方ヨリ（ナヒト ニヲト ゐヲ゜）突クナ 乙方ハ（ナヒト

應シテ突立左ノ右ニ足進ミ甲ノ右膝

ノ辺二五六寸前二踏止テ甲ノ手ノ

出ヲ又先ノ右ノ手ヲ矢筈ニシテ甲

ノ右臂ヲ唯ト押シ第一図ノ如シ

而シテ左ノ足ヲ甲ノ背後二踏出シ押ヘ兄依 後二廻リ

左ノ手ニテ甲ノ右肩ロ衣上ニ襟ノ辺ヲ猫ミテ左ノ肩

ロヨリ第二図ノ如ク襟肩ノ辺ヲ掴ミテ寸ト立殖ニ跂

後我ガ体ト共引付クルナリ甲方ハ此時右ノ拳ニテ図ノ如ク

乙ノ面ヲ打クントス乙ハ顔ヲ右ニ向ケテ外ス（頭姿 ）爪先見▲

▲双手ニテ甲ノ

襟首ノ所ヲジリ〳〵

トシメルナリ

甲方ハ手ヲウツテ

頌ヲ告ク

乙

甲

一図

乙

甲

二図

初段 Shodan: First Level Techniques
4/10袖車Sode Guruma: Sleeve Wheel

This technique begins the same way as True Stance, with both you and the attacker seated across from each other in Seiza with your hands on your knees.

Illustration 1

甲Koh
Attacker

乙Otsu
You

The attacker shouts *Eiya!* You respond with *Yaa!* and stands up. Starting with your left foot take two quick steps towards the attacker, stopping 5 or 6 Sun, 6~7 inches/15~18 centimeters, in front of his right knee. Before your opponent can react, form your right hand into a *Yahazu*, nock of an arrow, and shove him in the right upper arm. This is shown in the first illustration.

Next, step behind him with your left foot and reach over his right shoulder with left hand and grab his front collar. Then, with your right hand, reach over his left shoulder and grab near his left collar. This is shown in the second illustration.

Illustration 2
The small sketch on the bottom right is by a previous owner of this book

Raise yourself up slightly before stepping backwards, being sure to hold tightly so he will be yanked backwards.

As your opponent is pulled backwards, he will attempt to punch you in the face with his right fist. Avoid this by turning your face to the left and looking at the toes of your left foot. This is shown in the second illustration and is also similar to the second illustration of the first technique in the Shodan, First Level Technique, True Stance.

Steadily increase pressure on your opponent's neck. Your opponent should tap to show he is defeated.

Note: Adding the illustrations from the 1926 (illustrations 2 & 3) edition, as well as the sketch by an unknown previous owner of this book, makes the sequence easier to visualize.

1	2

3	4

5

飛違 トヒチガヒ

○此手ハ双方三尺ヲ隔テ少シ斜ニ向ヒ坐シ甲方ハ小太刀ヲ帯ヒテ搆ヘ甲ヨリ乙ノ真向ヲ見込(ヱヽト声掛小太刀ヲ引抜テ切付ル)乙方ハ(ヤット声合セ右手ヲ刀ニテシ甲切付ルニ腕ノ所ヲ押當唯ト受业メ左リノ手ニテ甲ノ手首ヲ第二図ノ如シ握リ図ノ如ク左ニテ取ル我カ左ノ腰ニ引付右ノ手ヲ矢筈ニナシテ甲ノ咽ヘ押當左ニ引付足ノ違立有リシヲ乙唯答ヘ甲ヲ廻シ投ニ倒シ甲右ノ脇ノ下ヘ我カ右ノ膝ヲ突キ右手ノ拇ニテ甲右ノ耳ノ下ヲ強ク押付左手ハ押ヘシ併我カ左ノ足ノ小指ヨリ甲ノ。

一図

二図

●握リタル柄ノ所ヘ指ヲ挿シ入レ二図ノ如クニシテ小太刀ヲ取ルナリ
甲ハ乙ノ右ノ肩ヲ引テ乙モ倒シ見ルベシ
乙モ我カ体ヲ搆ヘヨウニシテ倒サレヌヤウニズベシ

初段 Shodan: First Level Techniques
5/10 飛違 Tobi Chigai: Leaping In and Attacking (Reversing the Situation)

This technique begins with both combatants seated in Seiza 3 Shaku, 3 feet/90 centimeters, apart. The Attacker is slightly angled away from you and has a Kodachi, short sword, in his belt.

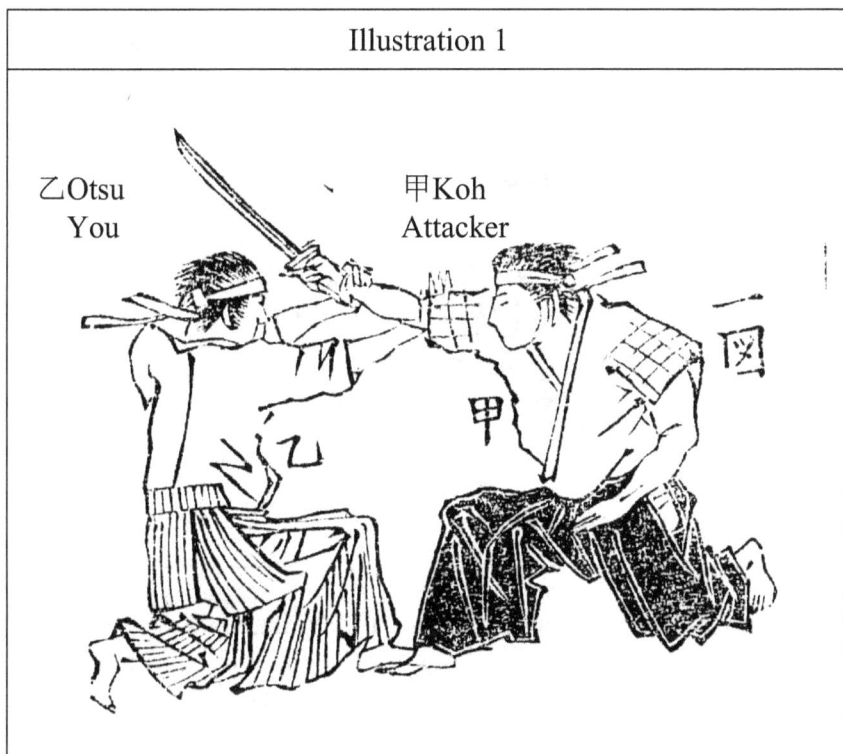

Illustration 1

The attacker turns his gaze on you and shouts *Eiya!* while drawing his sword and cutting straight down on top of your head. When attacking, he steps forward with his right foot and plants his left knee on the ground. He is cutting from a position known as *Nakagoshi*, or middle hips. You respond to the attack by shouting *Yaa!* and forming your right hand into a Shuto, knife hand, and striking his upper right arm.

Having stopped his arm, you seize your attacker's right wrist with your left hand. When striking his arm, you should step forward with your right foot and plant your left knee on the ground. This is all shown in the first illustration.

Illustration 2

甲Koh 　　　　乙Otsu
Attacker 　　　You

Next, while keeping a grip on his right wrist, yank your left hand down to your waist. Form your right hand into a *Yahazu*, nock of an arrow shape, grab your attacker's throat and push. The combination of pulling on his wrist and pushing on his throat allows you to topple your opponent to your left. As you do this, drop your right knee down beside his left ribcage and stand up on your left foot. Use the thumb of your right hand to press firmly into the vital spot just below your opponent's ear while holding his right wrist securely.

Use the toes of your left foot to push the sword handle out of his grip. This is shown in the second illustration.

After being thrown the attacker will attempt to drag you down by grabbing your right sleeve. Make sure you are well positioned in order to prevent this.

Illustration 1 & 2 from the 1893 version

一　遠　飛

甲Koh
Attacker

乙Otsu
You

二　遠　飛

甲Koh
Attacker

乙Otsu
You

拔身目附 ヌキミメツケ

○此手ハ雙方ノ構ヘ真ノ位ニ如クナリ甲方ハ

大太刀ヲ帶シ乙方ハ小太刀ヲ帶ニ二間餘ヲ隔

甲ヨリ壽樹ハ（ヲ乙方モ（ヤー）答ヘ小太刀ヲ引抜
甲ノ顔ヲ見込小太刀ヲ突出シサマ突キ上リ

右足ノ爪先ヲ甲ノ右膝五六寸前ニテ踏止メ（ヤート

言テ小太刀ヲシメシテ甲ノ目先ヘ突付ル

甲ハ（ヘイト言ヒ大太刀ヲ拔打ニ乙ノ足

ヲ拂ハントス乙ハ（ヤート其大太刀ヲ受止メ

直ニ大太刀ヲ打落（此ノ手音

突付ル故ニ甲ハ身ヲ左ヘ向ル因テ甲ノ背後（スキミ）ニ立テ甲ノ右ノ肩ロ

乙ヨリ我ノ左ノ手

ニテ甲ノ上ヨリ

取リ体ヲ下ゲ足

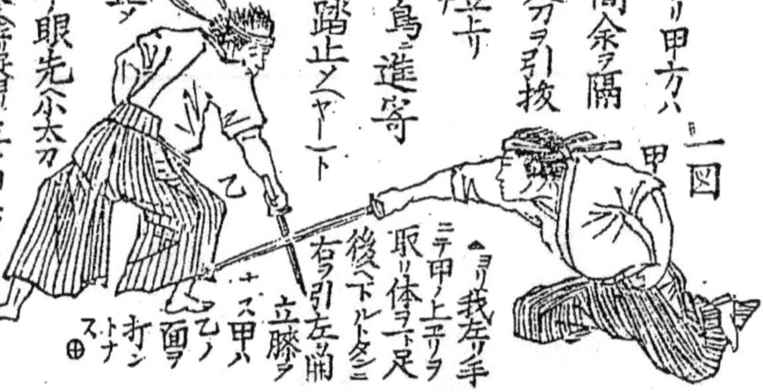

後ヘ下ルトキニ

右ヲ引ガ左ヲ腹

立膝ヲ

乙ハ甲ハ

タルマ

甲ノヱリト

首我ガ腕

ヲ押付ジリ〳〵メルナリ

甲畳ヲウツ

二戻千鳥ニ進寄

ヘ（但シ右ニ乙ハ（ヤー）言ニテ

ヲ二文字ニ構ヘ右膝ニ

二文字ノ如ク取ルハ上

襟ヲ掴ミ

二図 甲

立膝ヲ

初段 Shodan: First Level Techniques
6/10 抜身目付 Nukimi Metsuke: Locking Eyes and Drawing

Illustration 1 of Shin no Kurai, True Stance

乙Otsu
You

甲Koh
Attacker

This technique begins from the same position as shown in the first technique of the Shodan, True Stance. Both you and your opponent are seated in Seiza 1 Kan, 6 feet/ 1.8 meters, apart with your eyes locked.

The attacker has an *O-Dachi*, Long Sword, while you have a *Kodachi*, Short Sword, in your belt. Your opponent sits with his left hand on the scabbard of his sword while your hands are both resting on your knees.

Translator's Note:

The author uses *O-Dachi* and *Ko-Dachi* to describe the long and short swords worn by a Samurai. They can also be referred to as Daito, Long Sword, and Shoto, Short Sword.

Typically, "Tachi" describes the large swords that hung blade down from the hip by straps which Samurai mounted on horseback wore in the Warring States Era, however in the Edo Era 1600~1868, the long sword worn by a Samurai was also referred to as a Tachi. The 1926 version of this book gives the dimensions of the wooden practice sword: blade should be 3 Shaku and 2 Sun, or 3.2 feet/96 centimeters in length.

1. Tachi   – Long sword (worn blade down)
2. O-Dachi   – Long sword, more commonly known as Dai-to.
3. Ko-Dachi – Short sword, more commonly known as Sho-to.
4. Tanto   – Knife

	1 Shaku 1 ft/ 30 cm	2 Shaku 2 ft/ 60 cm	3 Shaku 3 ft/ 90 cm

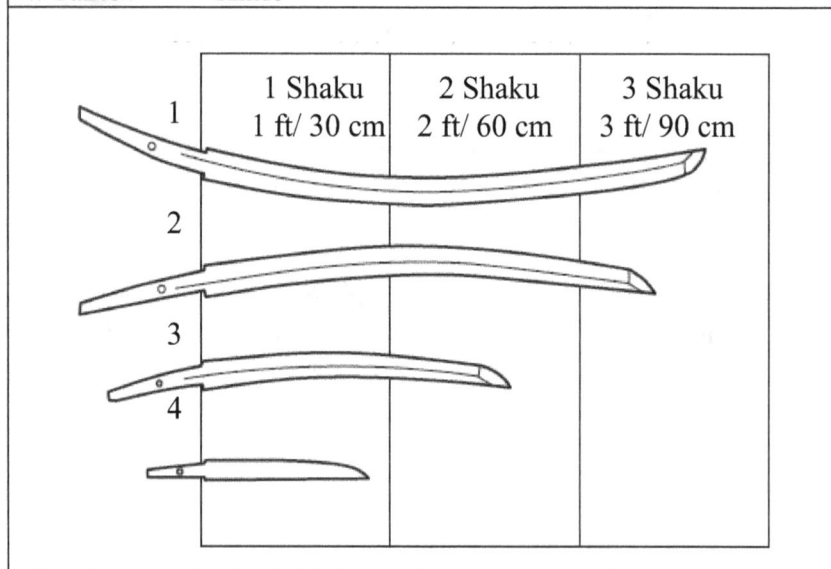

The technique starts when the attacker shouts *Eiya!* and you respond with *Ya!* while moving your right hand to the handle of your sword. When doing this, being careful not to grip too tightly. Draw your sword in one smooth, clean motion so you end up with your arm and sword forming a straight line from your shoulder to the *Kisaki*, tip of the blade. Aim the tip of your sword at your opponent's left eye.

The first illustration in the 1922 version shows the image after you draw your Ko-Dachi

一、
抜身目附一

Next, take two quick steps forward, like the Chidori bird, while keeping the tip of your sword aimed at his eye. Stop 5 or 6 Sun, 6~7 inches/15~18 centimeters, in front of the attacker's right knee.

Illustration 1

乙Otsu　　　　　甲Koh
You　　　　　　Attacker

Shout *Yaa!* and stab your short sword at your opponent's eyes. He will respond with a shout of *Eiya!* and, in one motion, draws his long sword and swings it, cutting across your right leg.

Respond with a shout of *Yaa!* and block his cut with your short sword. The best way to do this is to twist your hand so your pinky is up, thereby rotating the blade downward. After stopping the cut, knock his sword to the ground by deftly rotating the back of your hand clockwise in a quick motion.

Immediately return the tip of your sword to your opponent's eyes and thrust forward. This will cause him to flinch to his left. Use this chance to step behind him with your left foot.

Illustration 2

甲Koh
Attacker

乙Otsu
You

Next, reach over his right shoulder with your left hand and grab his collar then drop onto your left knee as you swing your right leg back, pulling your attacker down.

As your opponent falls, he will attempt to punch you in the face with his right fist. Respond with a shout of *Yaa!* and take Ichimonji Kamae, making your body into a straight line, and look at your right knee. This is shown in the second illustration.

With your left hand still holding your opponent's collar drop your left forearm into his neck and slowly increase pressure. He slaps the Tatami mat to indicate he is defeated.

○鎧返 コシリガヘシ

此ハ双方三尺余ヲ少
シ外シテ坐シテ構甲方大太
刀ヲ帯シテ甲ヨリ（ヤ）ト声ヲ掛ケ
乙ヲ抜打ニセントス乙ハ其気合ヲ
見テ（ヤ）ト應テ甲ノ柄ヲ左手ニテ押ヘ
右モ持添テ左足ヨリ立甲ノ左膝ノ
元（足ヲ踏出シ直甲ノ左横ニ付ト立
第一図ノ如ク右手ニ鎧ヲ取リ左手ニ甲ノ手首
ヲ取リ（ヤ）声ヲ掛ク我カ左リ足ヲ先ヘ四尺許左斜ニ引
付ヶ鼻ニ押伏ル但シ二図ノ如ク右膝ヲ甲ノ左リ脇ノ下迄ヘ突
キ左膝ヲ立テ鎧ヲ返シテ甲ノ上膊骨ニ押當テ手首ヲ捻ジテ正ニナシテ×

乙

甲

乙

一図

×ジデ〈ト押伏ゼルナリ
尢先ヲ見ル前ノ如シ
甲方痛ヲ覚ヘ
タミヨ打ツ

二図

甲

乙

初段 Shodan: First Level Techniques
7/10 鐺返 Kojiri Gaeshi: Reversing the End of Your Scabbard

This technique begins with both combatants seated facing slightly away from each other in Seiza 3 Shaku, 3feet/90 centimeters apart. Your opponent has long sword in his belt and is holding the scabbard with his left hand.

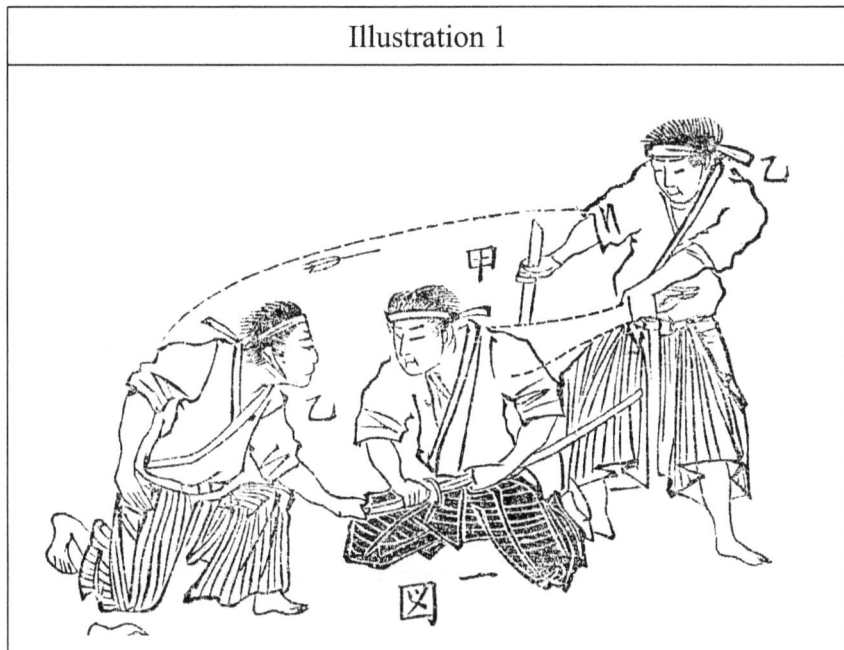

Illustration 1

The attacker shouts *Yaa!* and places his right hand on the handle of his sword, showing he intends to do a *Nuki-uchi*, draw and cut. Respond with a shout of *Yato!* and seize the handle of his sword with your left hand, stopping his draw. Stand up on your left foot while joining your right hand on the handle of your opponent's sword.

Next, step towards your opponent so you end up behind him near his left knee. Then, as the first illustration shows, seize the end of his scabbard with your right hand and grab his left wrist with your left hand. With a shout of *Yaa!* step diagonally to the left, starting with your left foot approximately 4 Shaku, 4 feet/1.2 meters, and yanking your attacker face down onto the Tatami mat.

Illustration 2
The bottom half of this illustration was partially erased, and then filled in by the previous owner of the book.

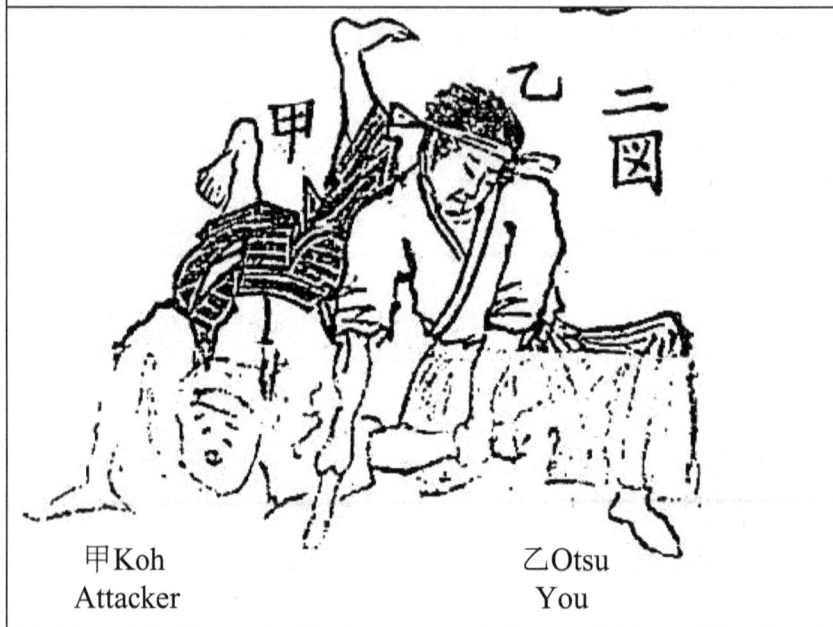

甲 Koh
Attacker

乙 Otsu
You

As Illustration 2 shows, plant your right knee beside the attacker's lower left side while keeping your left knee is upright. Force the end of his scabbard down so it presses into the back of his arm hard enough that you can feel the bone. You should bend his left wrist into a lock and gradually increase the pressure. Just as in the other techniques you should end by looking at your toes. The attacker should strike the ground when he feels pain.

Adding the illustrations from the 1898 edition (Illustrations 2 & 3)

1	2
	一　返　鎗 
3	**4**
二　返　鎗 	

両手捕　リヤウテドリ

○此手ハ双方共三尺

余ヲ隔向ヒ合膝三手ヲ

置キテ坐ス而シテ甲方

ヨリ(子イ)ト声ヲ揚ケ乙ノ両

手首ヲ取ル乙方ハ(エイ)ト應シ

其持レタル右ノ手ヲ前ノ

振解ノ如ク我カ左ノ肩先ニ引テ拂ハラヒ直ニ

甲ノ右肩ロノ衣類ヲ摑ミ又左手ヲ振掃ヒ甲ノ右手首ヲ取(乙ハ握リタル手ヲ乙ハ捻マルヽ必捨テ)

(エイヤ)ト言サマ甲ノ右膝ノ横ニ折ト立ツ此時ニ甲　右ノ肋骨ヲ一本跳込ム(但形ニ直)

ニ我が真前ニ引付茅二図ノ如ク捻伏セ右膝ヲ甲ノ上臂ニ押當テ左手ニ

取ルモ甲ノ手首ヲ一文字ニ立テ膝ニテジリ〱ト押付ルナリ

初段 Shodan: First Level Techniques
8/10両手捕Ryo-te Dori: Responding to a Two-Handed Attack

In this technique both you and the attacker are seated facing each other in Seiza with your hands on your knees.

乙Otsu   甲Koh
You   Attacker

The attacker shouts *Eiya!* and grabs both your wrists. You shout *Ei!* and free your right hand using the previously described Furi Hotoki. After yanking your right hand up to your left shoulder and thereby shaking off the attacker's grip, grab the fabric on his right shoulder.

Next, shake off the attacker's right hand which is gripping your left wrist and take hold of his right wrist. (Rotating your wrist around will allow you to easily free yourself from the attacker's grip.)

With a shout of *Eiya!* take a short step forward with your right foot so you end up beside the attacker's right knee. Next, kick the attacker in the ribs on his right side. (This is *Katachi* or just showing the motion without making contact.)

Note: While the text does not indicate which foot you kick with the 1893 & 1926 editions show the kick with the right foot.

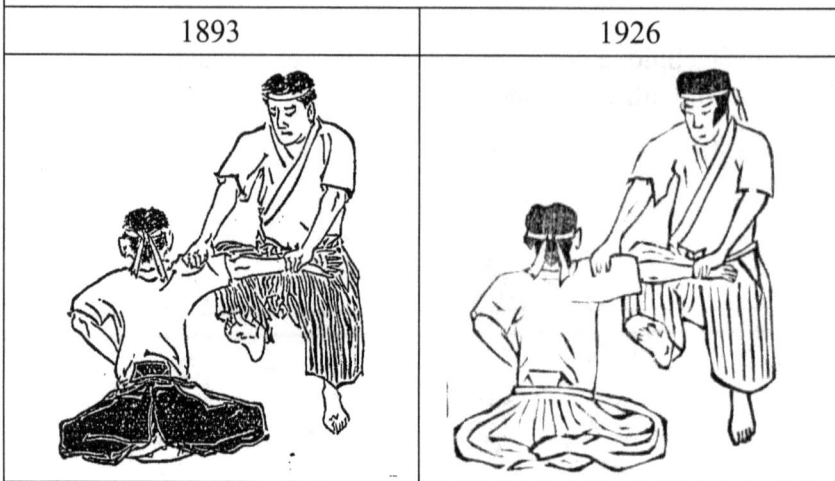

1893	1926

Illustration 2

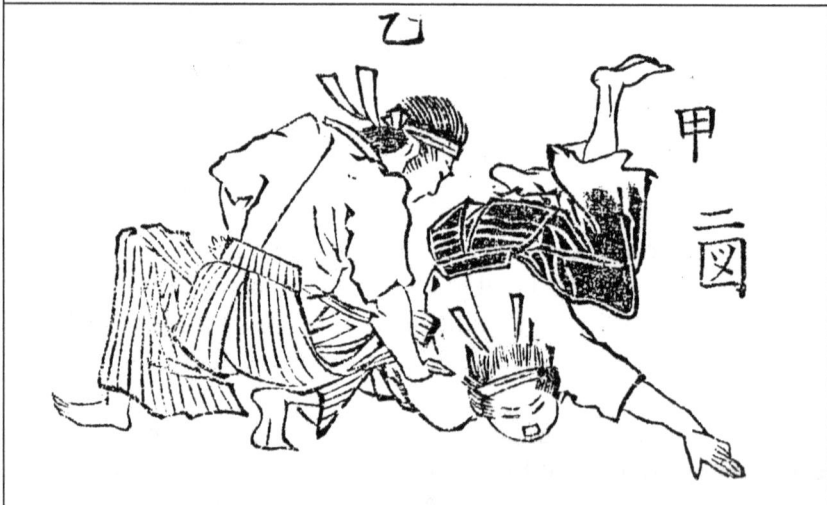

Then, as shown in the second illustration, twist and yank him down in front of you and plant your right knee on his right arm just above his elbow. With your left hand twist the opponent's right wrist so his hand is straight up so it makes an Ichi Monji, the Kanji for one, as you use your right knee to grind into the back of his arm.

Combination sequence with the 1887 and 1893 editions	
1	2
3	

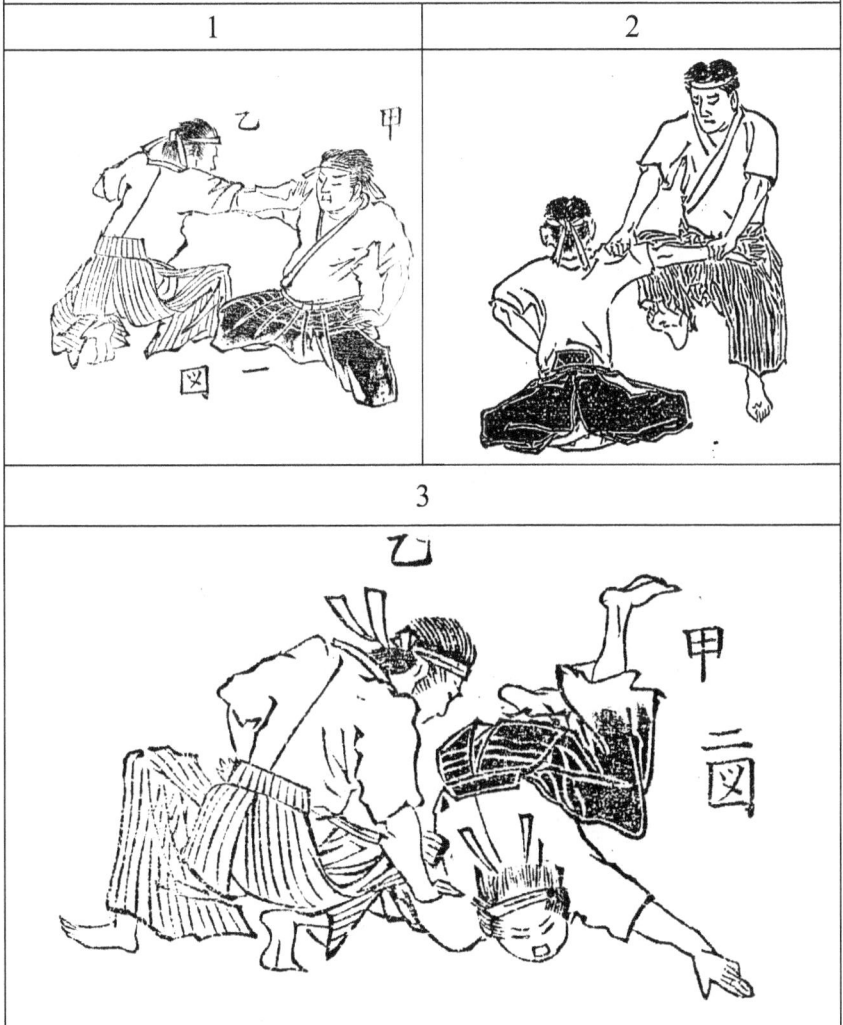

壁添 カベソイ

○此手ハ乙方ガ壁ニ背キビタリ付ニ我ガ両手ニテ翠ヲ圍ヒ乍右ノ足ノ膝ヲ立左リノ膝ヲ

コレニ向ヒ二尺四五寸前カ両手ニテ翠ヲ圍ヒ乍右ノ足ノ膝ヲ立左リノ膝ヲ

突第一圖ノ如クシ甲ヨリ（オヽ）ト掛声共ニ乙方ノ咽喉ヘ

右ノ手ヲ矢筈ニシテ押付左リノ手ニテ乙ノ

前帯ヲ掴ミ右ノ手ヲ向ヘ押ス左リハ前ヘ（引付ル

乙ハ（アイヤ）ト声掛ル途端ニ我カ腹ヲ前ヘ張出テ

我体ヲ延ハ両右ノ手ハ翠ヲ圍ヒ左リノ手ヲ矢筈ニシテ

我ガ頭上ヨリ甲ノ押付テ居ル右手ノ外ヨリ圖ノ如クシ

（ヤ）声ト共ニ我ガ腰ト矢筈ノ手ヲ同ニ下ルナリ此手矢

筈ノ手先ガ甲方ノ顋ノ辺（當テ直ニ其手ヲ掛ケ右ノ手ヲ甲頭ヘ捋添ヘ我ガ右ノ膝ノ上ヘ甲

ノ首ヲ引付捻廻シニ投ルナリ（乙投ト直ニ我翠ヲ圍フナリ）

口右ノ膝ヲ突キ左リヲ立膝ワナス等ノ口印ヲ付ス

甲

乙

初段 Shodan: First Level Techniques
9/10 壁添Kabe Soi: Pressed Against a Wall

This technique begins with you seated in Seiza with your back flat against a wall. Your arms hanging down and your hands are encircling your groin.

The attacker is seated 1.4 or 1.5 Shaku, 17~18 inches/42~45 centimeters across from you with his hands circling his groin.

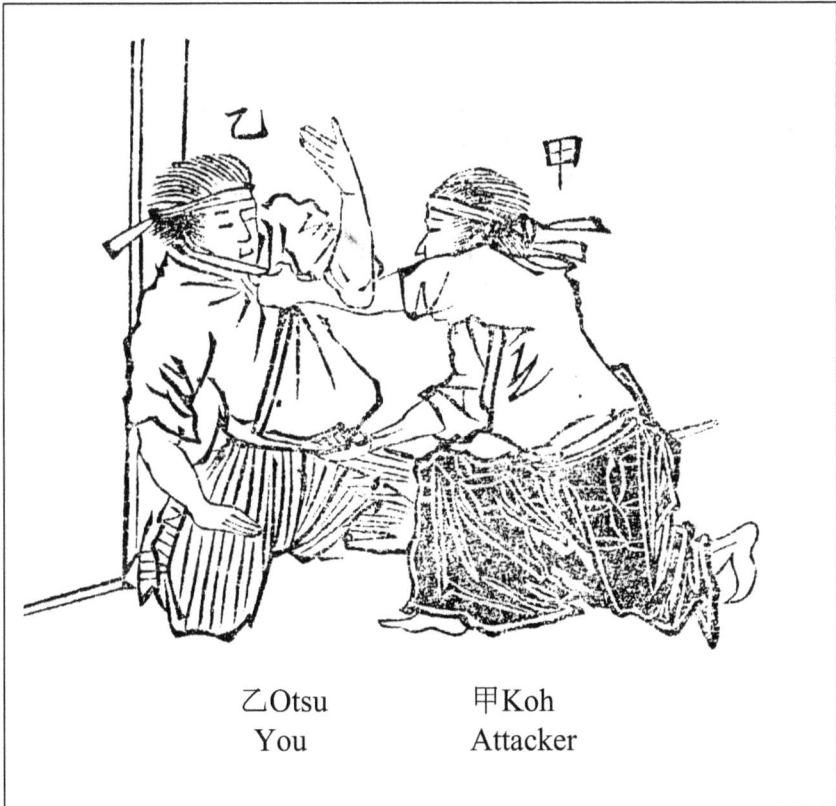

乙Otsu 甲Koh
You Attacker

With a shout of *Eiya!* the attacker steps forward with his right foot while keeping his weight on his left knee. He forms his left hand into a Yahazu, nock of an arrow and grabs your throat. With his right hand he grabs your Obi, belt, from below. He pushes back with his left hand and pulls forward with his right.

You respond with a shout of *Eiya!* as you extend your stomach forward while stretching your body upward. When doing this keep your right knee on the ground and stand up on your left foot. While keeping your right hand around your groin, raise your left arm over your head and form your left hand into a Yahazu, nock of an arrow shape. Swing your left arm across the top of your opponent's right arm as shown the illustration.

Keeping your left hand in a Yahazu, strike your opponent in the throat and then immediately grab hold of his throat. With your right hand grab his head. Twist and pull his head down towards your right knee. This will cause him to flip over. Once you finish the throw, return your hands to the start position, surrounding your groin.

Translator's Note:

The instructions in the 1887 edition state the attacker's right hand is grabbing the throat however the illustration clearly shows the left hand grabbing the throat. The instructions thus were altered to match the illustration.

While this book indicates the attacker has his left hand on your throat the 1893 and the 1926 edition (illustration below) describe the attacker's first move as being with the right hand.

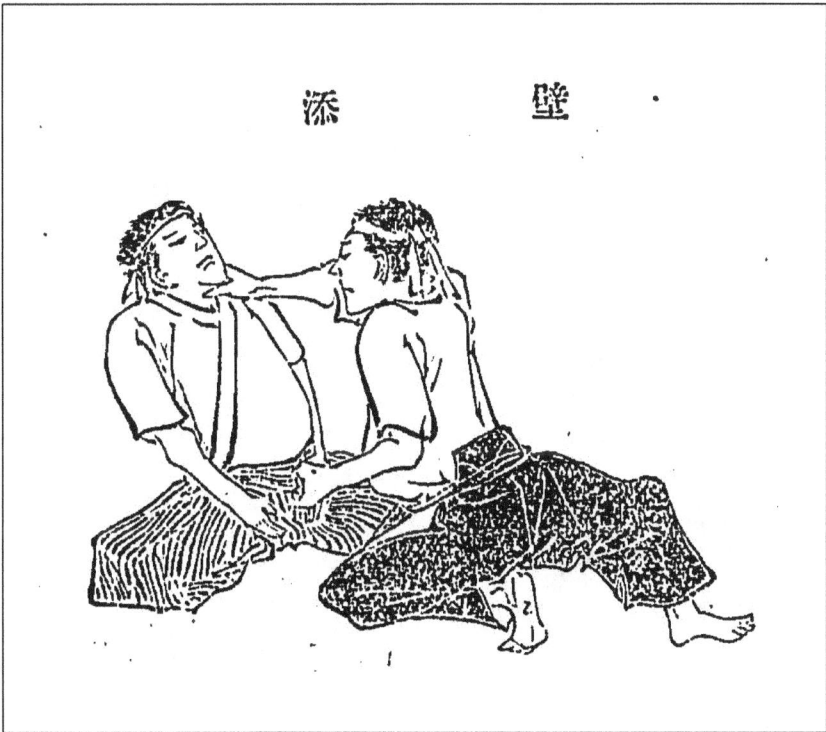

添 壁

後捕 ウシロドリ

○此手ハ乙方ガ壁際ヨリ三尺余ヲ隔向ヲ廣ク 一圖

明テ両手ニテ罘ヲ囲ヒ乍坐スナリ甲方ハ乙ノ後

ニ在リ（エイヤ）掛声ト共ニ左足ヲ立藤ヲナシ右ノ膝ヲ

突第一圖ノ如ク乙ノ背後へ抱付ナリ乙ハ（ヤ一）

答ノ声ト共ニ我ガ頭ノ頂ヲ甲ノ額へ当ル心ニシ

（咼顔ヲ居）我ガ躰ヲ少シ延上リ右ノ膝ヲ突左ヲ左ノ

後へ足ヲ開キ（ヤ一ト）声ヲ掛ルトタニ我ガ両肘ヲ張ルト

甲ノ抱メタル手ノ弛ム故直ニ我ガ腰ヲ下ルトタニ二圖ノ如ク

右ノ手ニテ甲ノ右ノ肩口ノ衣類ヲ掴ミ左ノ手ニテ甲足ノ

脚踝ヲ押ハネル（エイヤ）ト言テ我ガ前へ右ニ掴ミタル手ヲ強ク

引落ニ投ルナリ

甲

乙

二圖

乙

甲

一圖

初段 Shodan: First Level Techniques
10/10後捕 Ushiro Dori: Attacked from Behind

In this technique you are seated in Seiza, approximately 3 Shaku, 3 feet/90 centimeters, away from a wall with your arms ready and encircling your groin.

Illustration 1

乙 Otsu
You

甲 Koh
Attacker

The opponent, who is seated behind you, shouts of *Eiya!* as he presses his right knee into the ground, stands up on his left foot and grabs you from behind. This is shown in the first illustration.

You respond with a shout of *Yaa!* and whip your head back to strike him in the forehead with the back of your head. The person who is attacking should turn his head to the left to avoid being struck during training.

Next, extend your body upward slightly by putting your weight on your right knee, while swinging your left foot out to the left as you stand on it.

Illustration 2

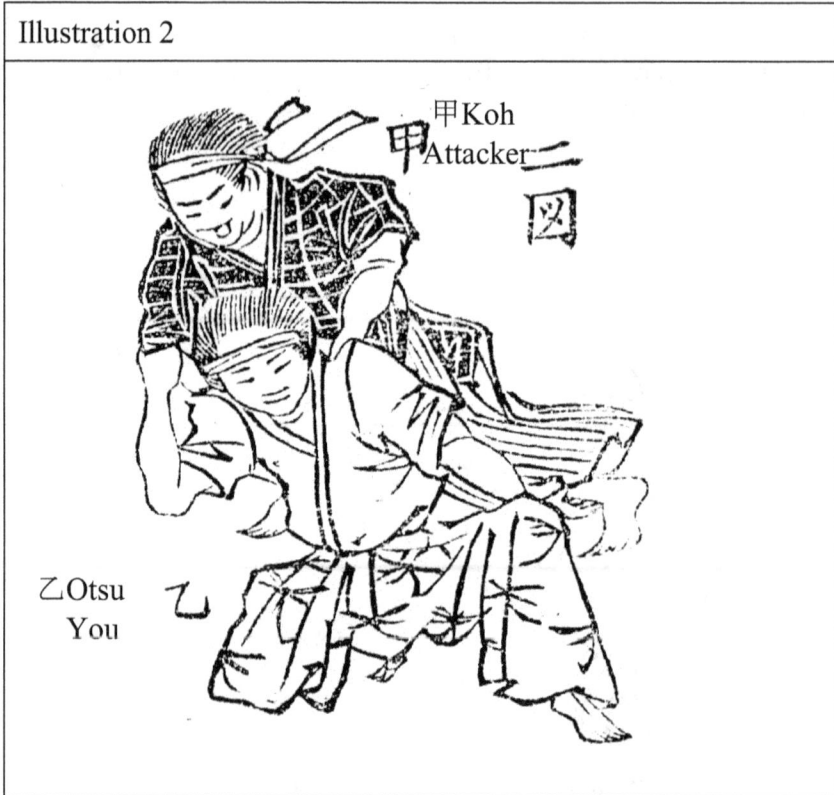

甲Koh
甲Attacker

乙Otsu
You

With a shout of *Yaa!* expand your elbows out in order to break your attacker's grip. As soon as you do this, grab the attacker's right shoulder with your right hand and push on the shin of his left foot with your left hand. As shown in the second illustration, throw by shouting *Eiya!* as you drop your hips and pull down hard with your right hand.

(After you finish the technique you should immediately return to a ready position, with your hands surrounding your groin and watching the attacker for any signs of a counter attack.)

(During training the attacker should be careful not to land hard on his right shoulder. You should be careful to throw the attacker lightly.)

Translator's Note :
This is a sequence that combines the 1887 and 1926 editions. The 1926 edition instructs you to stand up completely before dropping down onto your right knee.

1	2

3	4

初段立合十本ノ図

◯ 此手ハ双方共両手

行違 ユキチガヒ

際三立居構ヘテ甲方ヨリ
ニテ畢ヲ囲ミ乍各

（サヤ声ヲ掛乙方モ（ヤ下ト答ヘ双方

共中央ニ進出（既方共ニ員ヲ合セテ
声ヲ掛ル乙モ同ク答ヘ甲ノ右ノ手道ヲ左ニ手ニテ取擶右手ヲ手初ニシテ
甲ノ右肩先ヨリ腹ヘ搬（左ノ股ノ所ヘ斜ニ上ヨリ）手切入ル（此甲ノ体ヲ
一図如クニシ直又手首ヲ我カ左腰ニ引付右手ヲ甲右肩ヲ押
當右脚ニ甲ノ畢ヲ跳ト見セ此足ニテ甲右足ノ後ヲ掃フトダニシ

投ニ投リ我カ左ヨリ足ニ尻所ニ在踵ヲ車ザ廻ヘ（此ハ我カ左ヲ踵ヲ動カサベニ

甲

乙

一図

一寸 行違ト見セ甲ヨリ（ヤト

（甲ヲ投ヤイテ直ニ我カ右膝ヲ甲右ノ
脇下ノ所ヘ突左膝ヲ立右手ヲ捫
甲ノ右耳下ヘ押当ニ図ノ如ク
左ニ引分ニナスリ甲ハ左ノ
拳ニテ乙ノ面ヲ打ツ勢ヲナス、

（乙ハ員ヲ左リニ内ヘ此先ヲ見
ルベシ

乙

甲

二図

初段立合 *Shodan Tachiai* First Stage : Standing Attacks
1/10 行違 *Yuki Chigai* Crossing Paths

This technique begins with both combatants standing on opposite sides of the Dojo next to the wall. You should both be standing in Kamae with your hands circling your groin.

The technique begins with the attacker shouting *Ei-ya!* and advancing towards the center of the training area. You respond with an answering shout of *Yaa!* and move towards the center as well. (Both combatants should fix their eyes on each other and remember that it is important not to blink.)

The attacker makes it appear that he is going simply walk past you, when he suddenly shouts *Ya-toh!* You respond in the same manner and seize the attacker's right wrist with your left hand. You should raise it up slightly as you strike the attackers neck where it meets his shoulder with your right hand in a Shuto, Knife-hand.

Then yank the attacker's right arm diagonally downward towards your left thigh. (This will take your opponent's balance.) Then, as shown in Illustration 1, grip his right shoulder with your right hand and push down as you pull his right wrist to your hip.

Illustration 1

甲 一 図 甲 Koh Attacker

乙 Otsu You

With your right foot, feint as if you are going to kick the attacker in the groin, then use that leg to sweep his right leg. You are throwing him down with a *Mawashi-Nage*, Turning Throw.

When doing this throw, your left heel stays in place and you rotate on your left heel. (What this means is while your left heel stays in place, the rest of your boy is rotating and following your right foot.)

Immediately after you throw the attacker, plant you right knee by his right side while keeping your left knee up. Press the thumb of your right hand into the vital spot below the right ear.

Illustration 2

甲 Koh
Attacker

乙 Otsu
You

As shown in Illustration 2, you are pulling different parts of the attacker in opposite directions. His right arm is being pulled to the left while you press into the spot below his right ear, forcing him to the right.

The attacker will attempt to punch you in the face with his right hand. Avoid this by looking down at the toes of your left foot.

Translator's Note :

This book doesn't specify the Kamae, or Stance, used for this technique. Below are the fundamental stances from the 1926 version.

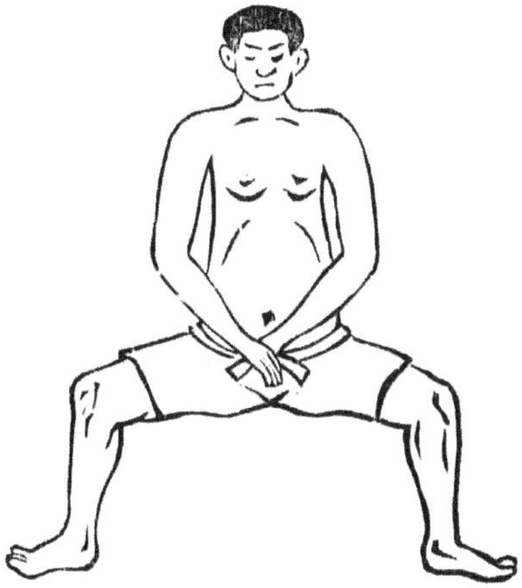 一、一文字形三圖	Ichimonji Kamae "One Line Stance"
 一、平一文字形四圖	Hira Ichimonji Kamae 2Open One Line Stance"

突掛 ツキカケ

○此手モ前ハ行違ノ如クナリ

双方共三掛声ヲシ中央マテ
進出三尺ヲ距(甲方ヨリ

声ヲ掛ケ(都テ立合三尺ヲ距レ見ル時
双方睾ヲ囲ヒ互ニ頁ヲ守リ油断
セヌ恐レリ(以下是皆)右拳ヲ以乙

左脇ヲ突此時右足
ヒ夕ヘ夕リ

下足進出シ乙ハ気見テ左肘ニテ脾腹ヲ囲ヒ右手ハ睾ヲ
囲ヒ(但シ左ハ手ハ半身ニ搆ヘ図ノ如)甲ノ突出シ
夕ル拳ヲ左リ手ニテ掴ミ右手ヲ持添(我カ体左リヘ廻シ

右膝ヲ突キ左膝ヲ立テ(ニイヤ)声ト共ニ投ゲタリ(但シ体ヲカハス
ミ(ノ時動作迅速)

甲ノ体ヲ我カ前ヘ二図ノ如ク引付甲ノ手先キヲ

一図
乙 甲

我カ臍ノ辺ヘ付テ腰ヲ
少シク延上リ前ヘ押出ス
ナリ(但シ此図ハ右掛リナリ
左モ此理ナリ
手ノ動作ハ前ニ示スカ如シ
甲ハ畳ヲオテ頂ヲ報ス

二図
甲
乙

突掛 *Tsuki Kake 2/10*
Attacked With a Punch

This technique begins the same way as the previously introduced Crossing Paths. You and the attacker both shout a Kake-goe and advance towards the center of the training area. When the attacker is about 3 Shaku, 90 centimeters, away he will shout as he steps forward with his right foot and punches with his right fist. His target is *Hidari Abara*, or the ribs on your left side.

(Be sure to note that before the two combatants are 3 Shaku apart, both should have their hands circling their groin, ready to defend themselves. Each should be watching the other carefully.)

You have anticipated this strike a pull your left elbow in, covering your ribs and abdomen, while your right hand protects your groin. (However, keep your left hand loose.)

Illustration 1

甲Koh
Attacker

乙 Otsu
You

At the same time step back with your right foot and open up to the right. This means you have shifted your stance so your body is perpendicular to the attacker. This is shown in Illustration One.

Illustration 2

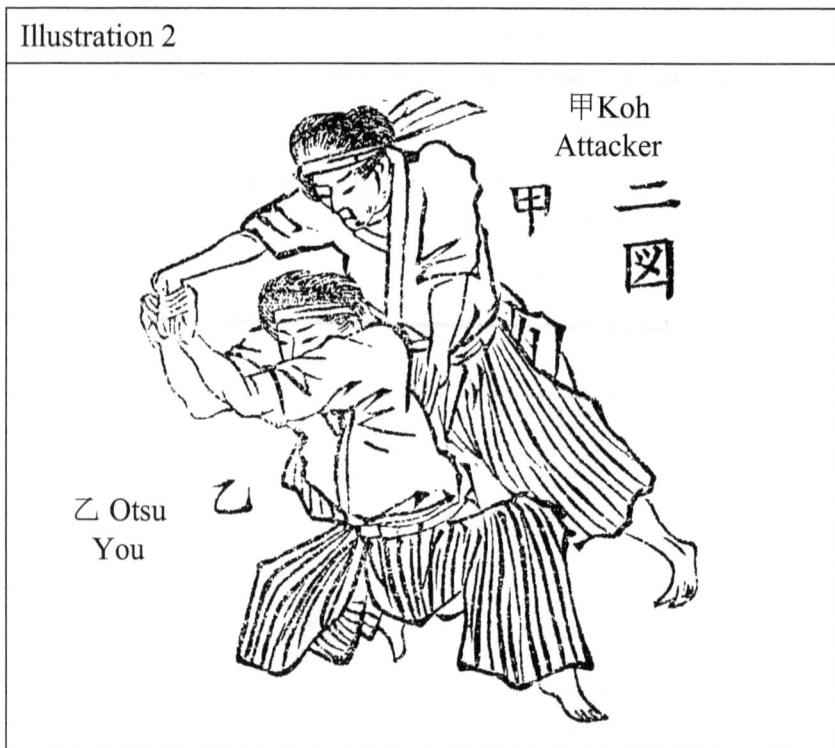

甲Koh
Attacker

甲 二図

乙 Otsu
You

乙

With your left hand seize the attacker's right hand that he punched you with. Then join your right hand beside your left. Rotate your body to the left counter-clockwise and drop down onto your right knee, keeping your left knee up. Throw the Attacker in front of you with a shout of *Eiya!*

(Remember that as you twist your body and drop down, you should increase your speed.)

As Illustration 2 shows, when you are throwing, yank the attacker's arms down to your navel. You should extend your back up and then forward as you throw.

(Note that these illustrations only show the right side, however the left side is done the same way.)

After being thrown, the attacker attempts to punch you with his left hand the same way as in the previous technique. The opponent should strike the Tatami mats to signal defeat.

引落 ヒキヲトシ

○此手モ前ニ同ジク

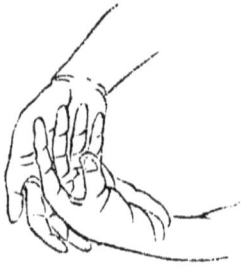

双方共ニ声ヲ掛ケ中央
迫進ミ三尺ヲ距リ甲方
ヨリ（ヤ）ト声共ニ右拳ヲ
以テ乙ノ脳ヲ見込ミ打掛
ル（左手ハ裏画）乙モ声ヲ合セ一図ノ如ク両手ヲ握リ（コブシ）

一図
甲

乙

〆甲ノ手先ヲ我ガ臍ノ辺ニ押シ
当テ左リ手ニテ甲ノ右肘ヲ押
付我ガ腹少シ前ニ出ス心得
三図ノ如ク押伏ルナリ
此手ハ少シ口傳アリ

甲ヨリ打込ム手ヲ十文字ニ受止メトタン右足ヲ飛ニ（トビ）
甲ノ拳ヲ跳込ム（祖形バガリ）受止メタル
右ノ手ハ甲ノ手先ヲ掴ミ左ハ甲ノ肘
二（エイ）声共ニ蹴タル足ヲ我ガ右後ニ開キ
押当（チヤ）声共ニ蹴タル足ヲ我ガ右後ニ開キ
手首我ガ右ノ腰ニ引付右足ヲ膝突左膝ヲ立甲ノ

二図
甲

乙

Shodan Tachiai First Stage : Standing Attacks 3/10
引落 *Hiki Otoshi* 3/10
Pull and Drop

This technique begins the same as the previous two. Both you and the Attacker shout a Kake-goe and advance towards the center of the training space. When you are about 3 Shaku, 3 feet/90 centimeters apart the attacker shouts *Eiya!* and punches straight at your head with his right hand.
(Note, that when punching, the attacker keeps his left hand surrounding his groin.)

You match the attacker's shout with one of your own and respond as shown in the illustration below.

Illustration 1

一図甲

甲 Koh
Attacker

乙 Otsu
You

Squeeze both hands into fists.
(This is called *Genko* 拳固 "fist + hard.")

Block and stop the Attacker's strike with a *Juji*, cross-shaped block. Let your right foot fly up and kick the Attacker in the groin.
(Note that for training purposes this is *Katachi Bakari*, or just miming the action of kicking.)

Next, your right hand, which you used to block, should seize the attacker's right wrist, while your left hand seizes his right elbow.

With a shout of *Eiya!* step back to the right with your right foot, which you just used to kick. As you do this, yank his wrist to your right hip. Your left knee should be up with you right knee on the ground. Keep the attacker's wrist against your stomach near your navel. Your left hand should be pushing his right elbow down.

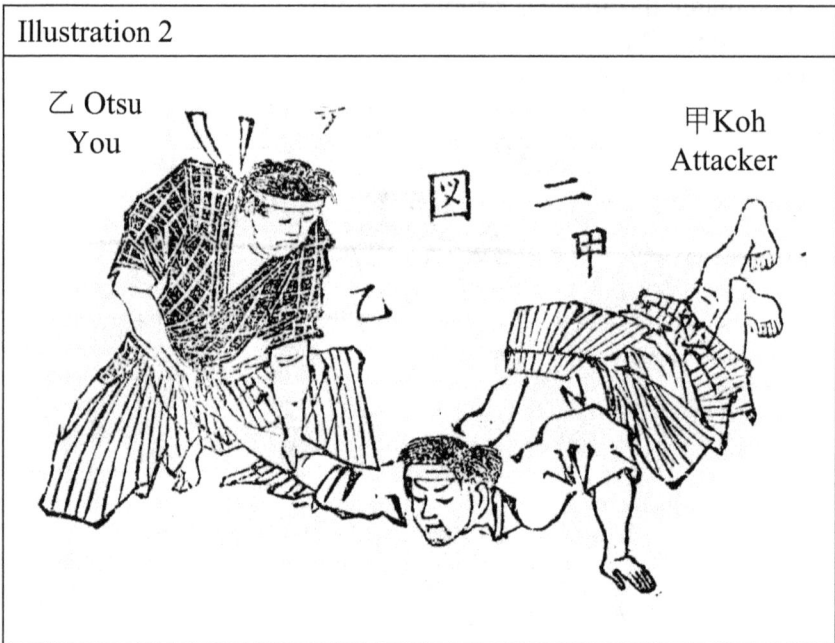

Illustration 2

乙 Otsu
You

甲 Koh
Attacker

Note that you should have your stomach pressed slightly forward. The Attacker should be brought down as shown in the second illustration. There are a few minor Kuden, orally transmitted secrets, about this technique.

Translator's Note:

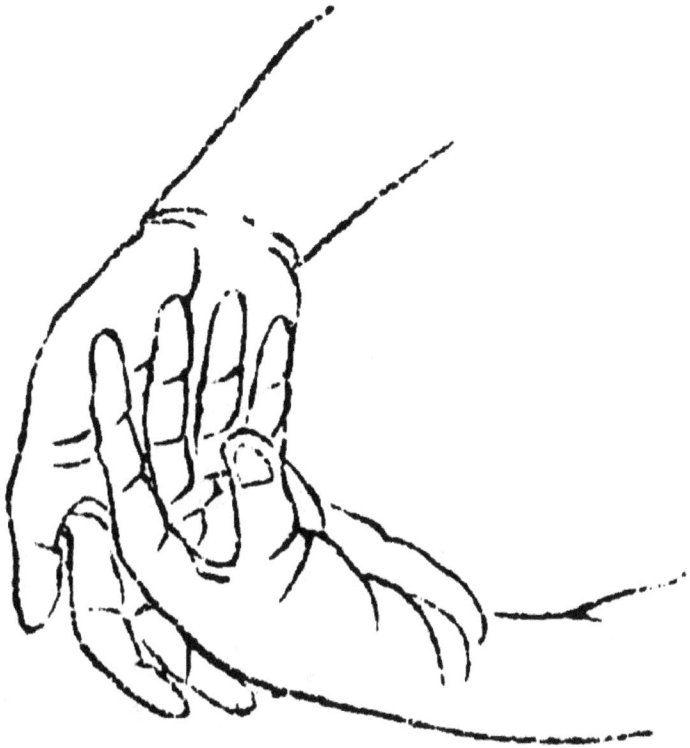

Translator's Note:
The text does not refer to the inset illustration of the hands, but it seems to highlight the final hand position.

両胸捕 リヤウムナドリ

○此手モ前ハ同シ

對シ甲方ヨリ(ヤ)ト

双方掛声ヲナシ中央

迫進ミ三尺ヲ距レテ

言テ乙方ノ襟ヲ取リ一図ノ如シ(但両手ニテ乙方モ又(ヤ)ト

第一図

声ヲ掛ケ甲ノ両手ヲ下ヨリ同シク両手ニテ甲ノ両

襟ヲ取リ(ヤエト)言サマ甲ヲ壁際マテ押付テ(ヤ)ト

声ヲ挙ケ我ガ左リノ膝ヲ突キ右ノ足先ニテ二図ノ如ク

甲ノ左股ヘ差入テハネ上ル其時我ガ体ヲ後ヘ吸返リ

持タル襟ヲ我ガ胸ヘ引付ル(但シ引ト(ヒネルト同時)甲方ハ成ル可ク

投ケラル、氣ニナリテ無理ナキヤウニスベシ

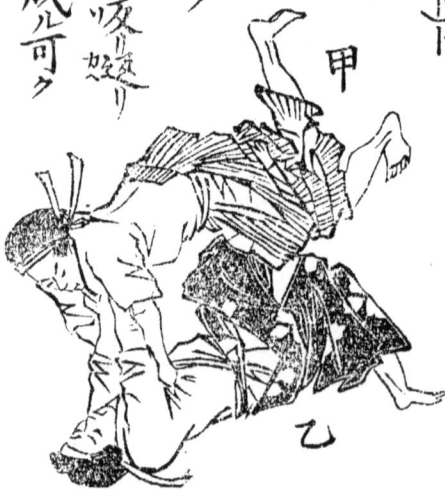

第二図

Shodan Tachiai First Stage : Standing Attacks
両胸捕 *Ryomune Dori* 4/10
Two Handed Chest Grab and Throw

This technique begins the same way as the previous one. Both you and the attacker shout a Kake-goe and advance towards the center of the training area.

You both stop when you are about 3 Shaku, 90 centimeters, apart. With a shout of *Ei!* the attacker grabs your lapels as shown in the first illustration. (Note that this is a two-handed grab.)

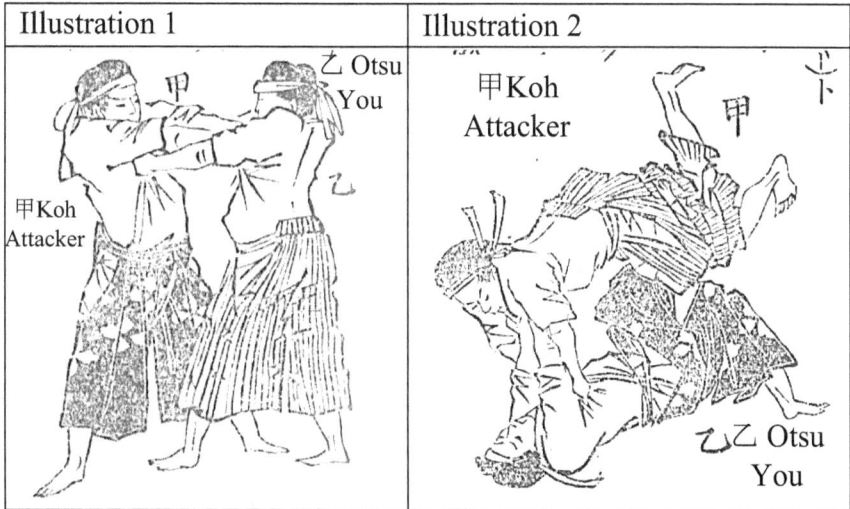

Illustration 1	Illustration 2

You respond with a shout of *Yaa!* and reach below the attacker's arms with both hands to seize his collar. Next, with a shout of *Eiya!* push the attacker until he is against the far wall. After that, shout a *Yaa!* and strike upward with your left knee. Slip your right leg forward until it is on the inside of the attacker's left thigh.

After that, throw your body down and backwards so you roll over while holding onto the attacker's lapels. Pull him towards your chest as you roll back.

(Note that you should be pulling and forcing him upwards at the same time.)

The attacker will find himself being thrown. Practice this carefully so you can throw without effort.

連柏子　（ツヒヤウシ）

○此手ハ双方共ニ壁際ニ並立双方

畢ヲカ丶ニ居リ甲方ヨリ（ヤー）ト声掛ケル

乙方モ之ニ應シテ共ニ中央マテ進

甲方ヨリ（エイヤ）ト言サマ抱付カントス乙モ

（ヤート）應シテ我ガ左リ手ヲ延シ甲ノ

左リ腰ニ手先ヲ掛ケ左リ足ヲ甲ノ右腿

ノ後ニ踏ニ張リ我ガ腰ヲ勦ク下ゲ図ノ

如クナシ（エイヤ）ト言サマ手ト腰トヲ捻ルトヲシニ張タル

左足踏耐ヘテ手ヲ後ヘ捻倒スナリ

此図ハ左リ掛リナリ左右共同理ニシテ乙甲ノ左ニ在時ハ左リ連捔子右ハ右

連柏子ト号ス

乙　甲

Shodan Tachiai First Stage : Standing Attacks
連拍子 *Renhyoshi* 5/10
Joined in the Same Rhythm

This technique begins with both combatants standing against the wall on opposite sides of the training space. Both are standing with hands circling the groin.

The attacker shouts *Ya!* and you respond with a shout of your own. You then both advance to the center of the training space.

The attacker steps past you, shouts *Eiya!* and wraps his arms around you from behind. You respond with a shout of *Ya!* while extending your left arm and gripping his left hip. Then shift your left leg behind the attacker's right thigh and drop your hips slightly. You should be positioned as shown in the illustration.

乙 Otsu
You
甲 Koh
Attacker

With a shout of *Eiya!* twist your arm and hips and put your weight on your left foot. As you twist you will topple your opponent over your leg.

This illustration only shows the technique on your left side. The principle applies to both the left and right sides. If the attacker is on your left side the technique is called *Left Joined in the Same Rhythm*. If it is done with the opponent on your right then it is called *Right Joined in the Same Rhythm*.

167

友車 トモクルマ

○此手ハ双

方片側ノ拵ノ

角三間余隔

テ向合ニ立居シ

乙方モ夢應ジテ其廳ヨリ三尺

許中央進ミ甲乙ノ巨離モ三尺許互ニ見合

テ気ヲ計リ甲ヨリ(ヤ)ト声掛ケ右拳ヲアゲテ乙ノ

眉涧(繋図)甲ヨリ(エ玉)

ヲ打ツ乙モ(ヤ)ト各一図ノ如ク右ノ手ヲ延シ正一文字ニ

受止ル(繋図)此時左ノ足ヲ左ノ後ニ引テ二尺余リ下ケ受止シ

手ニテ甲ノ手首ヲ取リ左ノ足ヲ甲ノ諸ヘ踏出シ右ノ膝ヲ突キ(一)

一 図 ナリ

甲
乙

○二図ノ如ク甲ノ手ヲ別搖我肘ヲ

脾腹ニ当(エ玉ト言サマ)領ヒ投ニスル
ヒハラ

此手ハ少シク入組ミタル

手故口傳アリテ六ケ敷ハ

熟覧シテ後取掛ルベシ

二 図

甲
乙

Shodan Tachiai First Stage : Standing Attacks
友車 *Tomo Guruma* 6/10
Roll Together

This technique begins with both combatants standing facing each other by one of the supporting columns approximately 2 Ken, 12 feet, apart on opposite sides of the training space. (Note that the hands should be surrounding the groin.)

The attacker shouts *Eiya!* and you respond with a shout of your own. You and the attacker then advance 3 Shaku, 3 feet/90 centimeters, towards the center of the training area. When the distance between you and the attacker closes to within 3 feet, you both lock eyes and judge each other's relative strengths and weaknesses.

With a shout of *Yaa!* the attacker raises his fist and punches to *Miken,* the spot between your eyebrows. You respond as shown in Illustration 1, by shouting *Yaa!* and raising your right arm up and blocking your opponent's attack. Your right arm should be held straight like *Ichimonji,* the Kanji for the number one, 一.
(Note that your left hand should be protecting your groin.)

Illustration 1

乙 Otsu
You

甲 Koh
Attacker

Illustration 2

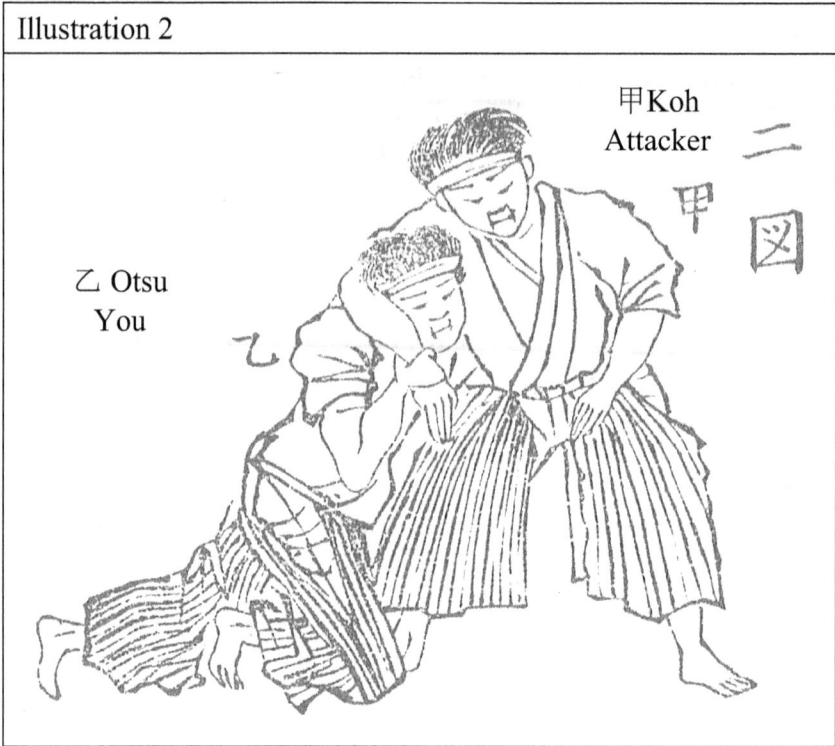

甲Koh
Attacker

乙 Otsu
You

二
図

甲

乙

At the same time, drop your left foot back and to the left about 2 Shaku, 2 feet/60 centimeters. Using the arm you blocked with, grab the attacker's right wrist. Then step behind the attacker with your left foot while dropping onto your right knee.

As Illustration 2 shows, wrap the attacker's right arm around your shoulders and strike him in *Hibara*, the side, with your left elbow. Shout *Eiya!* as you do this. Finally, do a *Shoi-nage*, load on the back and throw.

Since this technique involves getting in very close contact with your opponent there are *Kuden*, oral transmissions. Only train this difficult technique after you have studied the instructions carefully.

絹滞　キヌカツキ

○此手ヽ前同ク效方　一図　甲

壁際ニ立居（但シ翼ヲ画ク）甲

方ヲ（ニテ）ト声掛ヶ三四

尺程前ニ進ヱ方ヱ此

レ三答ヲ声ヲ掛ヶ甲ノ前ヘ進出（ナシ）（子ト）

言サマ甲ノ胸襟ヲ右手ニテ掴ミ左墓ヲ囲ヒ右膝ヲ突キ

左ノ足ハ左リ後ニ開キ立藤ヲナシ茅二図ノ如ノ甲

此時こガ取ル右ノ手首ヲ左リ手ニテ握リ（図ノ如ク下ヨリトル）

右拳ニテこノ䐣ヲ打ツハ我ガ取ル胸クラノ手

ノ下ヲカヒ潜リ右膝ヲ其侭ニテ廻リ左足ヲ左後ニ

開キ立藤ヲナシ（エイさ）言テ取ルタル襟ヲ前ヘ引キ（我墓ノ迎ヘ引）

（落ス心得ニスベシ）

一図
甲
乙

左リハ二図ノ如ク甲ノ左ノ銐
ヲハネテ投出スナリ

甲ハ投ラルヽトキ必右肩ヲ下
ケ我レヨリ先ニ向ヘ投ラル心
組ニスベシ

二図
甲
乙

Shodan Tachiai **First Stage : Standing Attacks**
絹潜 *Kinu Katsugi* 7/10
Dropping and Loading Like Silk[25]

This technique begins the same way as the previous one. Both combatants are standing against the wall on opposite sides of the training area. (Note that the hands are surrounding the groin, protecting it.)

The attacker shouts *Eiya!* and moves forward about 3 or 4 Shaku, 3~4 feet/90 ~ 120 centimeters. You respond with a shout and move forward. (Step forward about 3 feet.)

Next, with a shout of *Ei!* seize the attacker's left lapel with your right hand. Keep your left hand guarding your groin. After that, plant your right knee on the ground and swing your left leg behind you. This is shown in the first illustration.

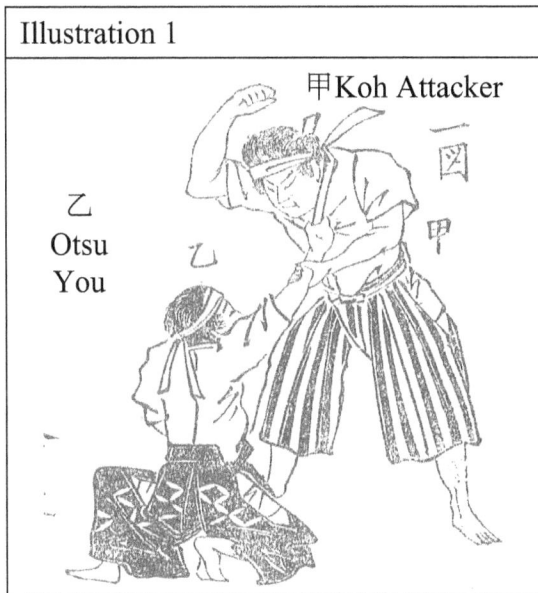

Illustration 1

甲Koh Attacker

乙 Otsu You

---

[25] Translator's Note:
The first Kanji means "silk" but the second Kanji means "to crawl or swim under something." The reading given for the second Kanji is Katsugi, which means "to load on the shoulder." I have included both meanings in the name of the technique.

Illustration 2

乙 Otsu
You

甲 Koh
Attacker

At this point the attacker seizes your right wrist, which is holding his chest lapel with his left hand.

(As the illustration shows, he grabs your wrist from below.)

The attacker then raises his right hand, readying to strike you in the head. While keeping your right knee on the ground and maintaining your hold on his lapel with your right hand, rotate your body counter-clockwise and extend your left foot back. Then, with shout of *Eiya!* pull his collar forward. (Note that your goal is to pull him down in front of your groin.)

As Illustration 2 shows, push back on his shin with your left hand as you pull with your right. The shin is called *Hagi* or *Tsune* in Japanese. This will throw your opponent down in front of you.

襟投 エリなげ

○双方前ノ如ク

畢ヲ囲ヒ左壁

際立居之甲方ヨリ

（ヱイト声掛ケ乙方モ

答ヘテ共ニ中央へ

進ミ摺リ違フテ行スキ

セ乙ハ甲ノ後襟ヲ一図ノ如ク取リ

我カ体ノ向キヲカヘヱリヲ取タル手ノ下ヲ潜リ左

膝ヲ突キ右ヲ立（左手ハ）（畢ヲ画）（ヤート声掛ケ甲ノ襟ヲ

我カ前ヘ引落ス（我ガ畢）（ニ引ク）二図如クナス

甲ハ倒レナカラ直ニ右手ヲ（手ガニナ）以テ乙ノ貝ヲ打×

一図

乙

甲

×乙ハ投ルヽイナ甲我ガ面ヲ

打ッ故ニ同シク手刀ヲ額

ニ加ヘテ受止ルナリ

二図

甲

乙

Shodan Tachiai First Stage : Standing Attacks
襟投 *Eri Nage* 8/10
Collar Throw

This technique begins the same as the previous ones, with both combatants standing against the wall on opposite sides of the training space. The hands should be protecting the groin.

The attacker *Ei!* and you answer with a shout of your own. Next, you both advance towards the center of the training area. As shown in Illustration 1, just as you pass the attacker, turn around and grab the back of his collar with your right hand.

Illustration 1

You then turn so you are facing the opposite way. Drop down onto your left knee directly under the hand you grabbed his collar with while keeping your right leg upright.

(Note that your left hand should be protecting your groin.)

Illustration 2

甲Koh
Attacker

乙 Otsu
You

As shown in Illustration 2, with a shout of *Yaa!* pull him down. (Your goal is to make him drop in the area near your groin.)
As soon as the attacker hits the ground he will try and punch you in the face with his right hand. (He will use a Shuto, knife hand.)

Since you are expecting this attack to your face, the moment you finish throwing you also make a Shuto and block the attacker's strike to your forehead.

手髪捕 タブサドリ

○此手ハ双方

向合三間隔

壁際ニ立（双方恐ヲ）

（ア図ニ甲方ハ左ト）

寿掛ケ乙（ヤイト答）

共中央ニ進ミ三尺ヲ隔テ止リ

図 一

甲

乙

甲方（ヤイト言ザ行成リ）乙ノ手髪ヲ右手ニテ掴ム
乙（ヤイ）と答ヘ甲掴ミタ手ノ外ノ掌上ヨリ握固如

（但此同時左ヲ足ヲ尺）

（ア図ニ如如如ス）復左足ヲ甲ノ右足元

マテ踏出シ同時左ノ手ヲ甲ノ肘ニ当テ図ノ如ク如シ

（乙ノ首ヲ抱ラヘメル）右ノ手ニテ握名甲ノ手ヲ我ガ小指ヨリ差入手髪ヲ放シ直ニ右ノ腰ノ辺へ引付
（故ニ手頭間冬）

見

甲

（ア伝）我ガ右足ノ際へ引
伏セ甲右脇下ニ左
膝ヲ半ヲ立テ此先ヲ

ヽ甲ノ体ヲ我ガ左リ腰ニテ崩シ（此甲ノ右）
肘ヲ我ガ左ノ掌ヲ当テ取リタル手先
ヲメ上ケ図ノ如クニテ向前ニ捻ル

二図

乙

甲

Shodan Tachiai First Stage : Standing Attacks
手髪捕 *Tabusa Dori* 9/10
Escaping From Someone Grabbing Your Hair

This technique begins with both you and the attacker standing against the wall on opposite sides of the Dojo about 2 or 3 Ken, 12~18 feet/3.6 ~ 5.4 meters, apart. (The hands of both combatants should be protecting the groin.)

The attacker shouts *Ei!* and you answer with *Ya!* and you both advance towards the center and stop when you are about 3 Shaku 3 feet/90 centimeters apart.

The attacker then shouts *Eiya!* steps forward and grabs your hair with his right hand. You respond with a shout of *Yaa!* and place your hand on top of his and grip. This is shown in the illustration.
(Note that at the same time you drop your left foot back about 2 Shaku, 2 feet/60 centimeters.)

With a shout of *Eiya!* step forward with your left foot so you end up beside the attacker's right foot. At the same time, strike and grip his right elbow with your left hand. You should be positioned as shown in the first illustration.

甲 Koh
Attacker

乙
Otsu
You

Next, drop your hips slightly.

(If you bend your head forward you will make a gap between the Attacker's hand and your head.)

You have seized the Attacker's right hand with your right hand. Start working the fingers of your right hand underneath the Attacker's hand, starting with your little finger. As soon as you free your hair, break his balance by pulling his right arm down beside your right hip forcing the Attacker's body beside your left hip.

(You are able to do this since you have the Attacker's elbow in the palm of your left hand and the end of his right hand in your right hand. This is shown in the illustration. You are twisting him forward. There is a Kuden, oral transmission, regarding this point.)

乙
Otsu
You

甲 Koh
Attacker

Pull the Attacker down beside your right foot. Plant your left knee in his right side, below his armpit, while keeping your right knee upright. Keep your eyes focused on the tips of your toes.

This combines the illustrations from this book along with the ones in *Illustrated Guide to the Inner Secrets of Tenshin School* by Kushi Niju 大串仁十, published in 1926. The title of the technique is the same but there are some variations since this is from a different line of Tenjin Shinyo Ryu Jujutsu.

1	2
3	**4**

後捕 ウシロトリ

○此手ハ前ノ居捕ノ後取ノ通リナリ

只立合ノ仕組ナリ乙方壁際ヨリ四尺

斗リ隔テ両手ニ峯ヲ囲ヒテ立ツ甲方其

後ヨリ（エイセ）ト声掛ケ乙ノ背ヨリ抱ク

第一図ノ如クナリ乙モ（ヤー）ト答テ前

ノ如ク我カ背頭額ヲ甲ノ顔ヘ（顔ヲ左ヘ向ケ）

一寸当直ニ我ガ両肘ヲ張ル（此ヲ伸ノ抱タル）

前ノ居捕ノ如キ手ニテ体ヲ下ゲ（両手ヲスカシ伸ヲ技クベシ）

右ノ膝ヲ突キ左足左ノ後ニ開キ（エイサ）ト言サマ

甲ヲ前ヘ引キ投ルナリ甲ハ投ラレバ乙ノ向ヘ

巳ヨリ先ヘカヘル心ニテ我額ガ乙ノ臍ニ当ル位ニ心得ベシ

Shodan Tachiai First Stage : Standing Attacks
後捕 Ushiro Dori 10/10
Grabbed From Behind

This technique is the same as the earlier technique from Idori, Seated Techniques, which is also called Ushiro Dori. However, since this section is Tachi Ai, Standing Techniques, it is done while standing. Other than that, it is done the same as the earlier technique.

You begin this technique standing near a wall with your hands surrounding your groin. The Attacker stands about 4 feet behind you. He shouts *Eiya!* and then begins approaching you from behind. As shown in Illustration 1, when he reaches you, he wraps both arms around you from behind.

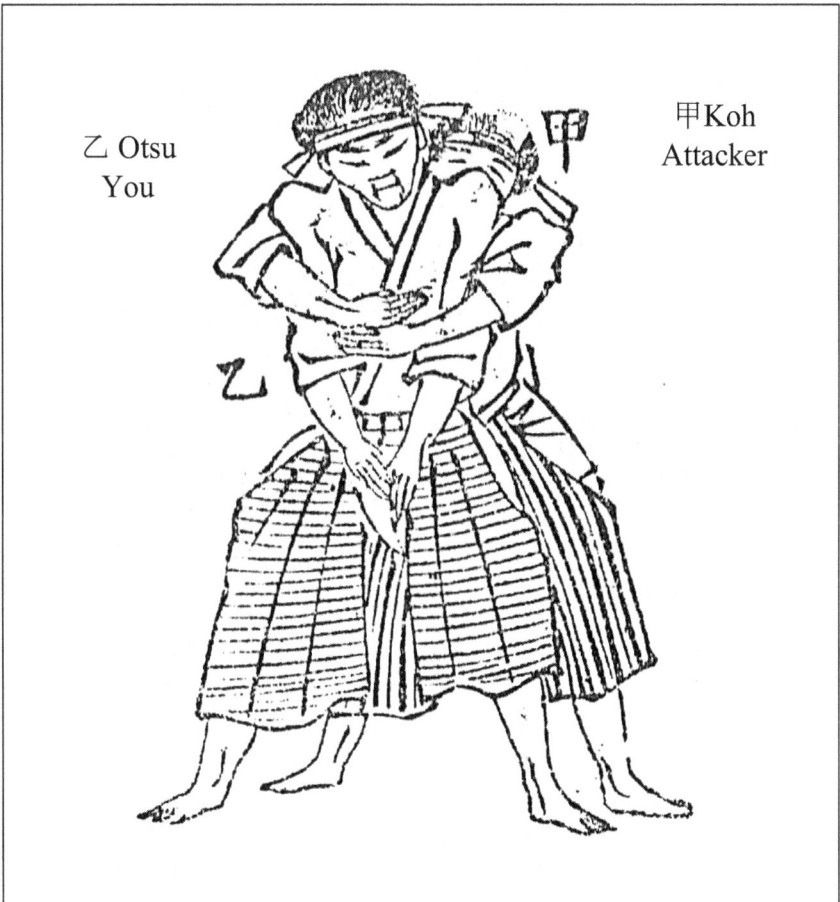

乙 Otsu
You

甲 Koh
Attacker

You shout *Ya!* and respond as you did in the Ushiro Dori in the Seated Techniques. Swing your head back so you hit the Attacker in the face. (Your attacker should turn his head to the left to avoid getting hit with the back of your head.) At the same time, push your elbows out to the sides. (This will enable you to escape his arms.)

Next, use your hands in the same way as described in the Seated Technique version of this technique. Drop down and plant your right knee on the ground and pull your left foot back. With a shout of *Eiya!* pull down and thrown him forward.

The attacker should be prepared to be thrown forward. He should end up with his forehead near your navel.

○此手ハ双方共前

投捨 二十本 撞木 シユモク

ノ初段立合ニ如

心得ヘシ陳テ双方

三間隔テ壁際

立（両手ニテ）甲方リ

（ヤヤ上声）掛中央テ

進ミ又同シ三尺隔テ止リ双方共身ヲ見合ヲ

都而投捨ハ形ヲ捕初ヨリ終リマテ双方

（共ニ白眼合テメダシキセス心エ也）甲ヨリ（エイト声

掛ケ右拳ヲ以テ乙ノ眉間ヲ打ントス乙方ハ左ノ手

ノ指ヲ延シ正一文字ニ図ノ如ク受止メ直ニ其手首ヲ

握リ右手ハ罪ヲ囲ヒ右ノ足ヲ右後ニ二尺許下リ（エイヤト言ヒ其取タル手首

我ガ左ノ腰ニ引付ケ又右ノ手ヲ

甲ノ左ノ肩ニ押当テ右足ヲ

甲ノ右ノ腿ノ後ヲ拂フ（三拍子合ス

分ケヲ（ケ合ン）我ガ左ノ腿ヲ其侭廻シテ

躰中腰ニ構ヘ両手ニ罪ヲ囲ヒ甲ノ再

ヒ掛ランヲ以テ注意ヲナス

初段投捨 *Shodan Nagesute*
First Level Throwing Techniques
撞木 *Shumoku* 1/20
Piercing Tree

The reader should understand that these techniques start out the same way as the previous Shodan Tachi-ai, or *First Level Responding to Armed and Unarmed Attacks.*

The first technique, Piercing Tree, begins with both combatants standing against the wall on the opposite sides of the Dojo, about 12~18 feet apart. (Their hands should be covering the groin.)

The attacker shouts *Eiya!* and advances towards the center of the Dojo. You do the same and you stop about 3 feet from each other. You should both be glaring fiercely at each other so that it seems your eyes have gone white.

(When doing Nagesute, Throwing Techniques, it is important for both combatants to maintain Hakugan, a cold glare, from the beginning to the end of each technique.)

乙 Otsu
You

甲 Koh
Attacker

The attacker shouts *Eiya!* and strikes to your Miken, between the eyebrows. You block this as shown in Illustration 1, by raising your left arm in the shape of Ichimonji, like the Kanji for one 一, while extending the fingers of your left hand. After blocking, use that same hand to immediately grab his right wrist, while being sure to cover your groin with your right hand.

甲Koh
Attacker

乙 Otsu
You

Next, step about 2 feet behind the attacker on his right side. With a shout of *Eiya!* pull his right wrist down to your left hip. At the same time shove his left shoulder with your right palm and sweep his right calf back using your right leg. (This is done in three beats. Starting with the Kiai of *Eiya!* and finishing when you sweep his leg.)

Note that when rotating you should have all your weight on your left heel and allow your back to bend. After the throw, both hands should be encircling the groin as you maintain a vigilant watch, ensuring the attacker does not strike again.

対捨（カリステ）

○此手モ前ハ双方

共ニ同ジク甲乙声ヲ

合テ中央ニ進ミテ

三尺ヲ隔テ止マ甲

方ヨリ（ヱイ）ト声ヲ掛

右ノ拳ヲ上ゲ乙ノ脳

部ヲ打タントス（左リ手ハ拳ヲ曲）乙方一図ノ如ク

右手ノ指先ヲ延正一文字ニ受此ノ（左ニ手ハ拳ハ）

真ニ其手首ヲ取ヱ左リ足ヲ左リ後ヘ

二尺斗リ下リ（ヱイ）ト声掛ケニ図ノ如ク甲ノ手首ヲ

右ノ乳ノ辺ヘ引付仍ヲ一文字ニ攞ヘ甲ノ左ノ肩ヘ左ノ手ヲ掛ケ▲

一図

甲

乙

▲左ノ足ハ甲ノ股ノ後ニ立少シ

腰ヲ下ゲ（ヱイ）ト言サマ我ガ真後

ニ投ルナリ以後乙ノ攞ヘ前同

甲ハ畳ヲ打テ負ヲ示ス

二図

乙

甲

Shodan Nagesute First Level Throwing Techniques
苅捨 *Karisute* 2/20
Throw Like a Scythe Through Grass

甲Koh
Attacker

乙 Otsu
You

This technique begins the same way as the previous with both you and the attacker shouting in unison. You both move towards the center of the training area and stop about 3 feet apart. Your opponent shouts *Eiya!* raises his right fist and swings, aiming to strike you in the head. (While doing this he keeps his left hand covering his groin.)

You respond as shown in the first illustration. Block his strike with your right arm in Ichimonji, straight like the Kanji for one 一, while extending the fingers of your right hand. (Your right hand should be covering your groin.) Then immediately grab that wrist.

乙 Otsu
You

二
図

甲 Koh
Attacker

Next, step about 2 feet behind the attacker with your left foot and, with a shout of *Eiya!* pull the attacker's wrist towards your right breast, straightening his arm as if it is in Ichimonji Kamae. This is shown in illustration 2.

Seize the attacker's left shoulder with your left hand. Press your left leg up against the back of his right thigh, lower your hips slightly and with a shout of *Ya!* throw him directly behind you.

Finally, return to Kamae as before and the attacker strikes the Tatami mat with his hand, indicating he has been defeated.

朽木倒 クチキタフシ

○此手モ、初ハ同シク双方掛声ヲ合セ

共ニ中央マテ進ミ三尺許ヲ隔テ止リ

双方顔ヲ見合ヒ氣ヲ窺イ甲方ヨリ（ヱイ卜

言サマ右拳ヲ以テ乙ノ眉間ヲ打

タントス、此時左ノ手ハ睪ヲ囲ト右ノ

足ヲ前ヘ一足踏ミ出ス図ノ如乙方

モ（ヤーイ卜應シ甲ノ拳ヲ振上ル時我が

右ノ足ヲ足進メ右手ノ外掌ニテ甲

胸ヲ押左ノ手ニテ甲ノ踏ミ出シタル

左ノ腿ヲスクヒ上（アイヤ卜言サマ向ヘ突倒ス

甲ハ倒レテ変起上リ掛ラノ勢ヲナス乙ハ両手ニテ睪囲ト右ノ足ヲ踏出タル侭身構ヲ

Shodan Nagesute First Level Throwing Techniques
朽木倒 *Kuchiki Taoshi* 3/20
Toppling a Decayed Tree

乙 Otsu
You

甲Koh
Attacker

This technique begins the same way as the previous with both you and the attacker shouting in unison and moving towards the center of the training area. Both of you stop about 3 feet apart. You should each be looking into the other's face, judging your opponent's fighting capacity.

The attacker shouts *Eiya!* and raises his right hand, ready to strike you in Miken, between the eyebrows. As the attacker steps forward with his right foot, he keeps his left hand covering his groin.

You respond by shouting *Yaa!* and stepping forward with your right foot the moment the attacker raises his fist. Shove him in the chest with the palm of your right hand. Use your left hand to scoop up his right thigh. With a shout of *Eiya!* throw him violently back.

The attacker may get right back up and resume his assault so you should stand ready, with your right leg forward and have both hands covering your groin.

横車 コシクルマ

○此手モ最初ハ前ト同レク双方声掛ケ互ニ

畢ヲ圍ト作シ中央マテ進ミ寄三尺ヲ隔テ、

甲方ヨリ（エイヤ）ト声ヲ掛ケ右足ヲト足

踏出シ直ニ両手ニテ乙方ノ両腰帯ヲ

取組付ヲ乙方ハ掛声ニ應シ左手ハ

畢ヲ圍ヒ右ノ手ハ指先ヲ延シテ図ノ

如クナシ甲ノ左リ腿ノ後ヘ（我右足

ヲ踏張リ身ヲ半身ニ構ヘ（エイヤ）ト

言ヒサマ甲ノ水落ノ辺ヘ右ノ肘ヲ上ヨリ

打落スナリ　（肘ヲ以テ胴ノツカ形ニ
　　　　　　　　活法ニ不熟ニシテ強ク覚悲）（甲ハ都而乙ノ気ニ合テ突掛蹴
　　　　　　　　　　　　　　　　　　込シ當テ并ニ強ク當ルヤウニスベシ）

甲
乙

Shodan Nagesute First Level Throwing Techniques
腰車 *Koshi Guruma* 4/20
Tossing Over Your Hip Like a Wheel

甲

甲Koh
Attacker

乙

乙 Otsu
You

This technique also begins the same way as the previous with both you and the attacker shouting in unison and moving towards the center of the training area, hands protecting the groin, and stop 3 feet apart.

The attacker shouts *Eiya!* and grabs both sides of your belt, holding you fast. You respond with your own shout and extend the fingers of your right hand while keeping your left hand protecting your groin. Then, as shown in the illustration, move your right leg and place it behind the attacker's left thigh.

With a shout of *Eiya!* slam your right elbow into his Mizo Ochi, solar plexus. (Your elbow should fit into the hollow formed where the diaphragm is located, but this should not be a true strike, just showing the form. Since you are not well versed in resuscitation, striking this point hard is a bad idea.)

(The Attacker should pay attention to how the defender is moving and be sure not to hit, strike or kick him too hard.)

一横　車　ヨコグルマ

此手合ノ形ハ中三間計リ離レテ

直立ナシ互ニ聲ヲ懸テ進ミ中

三尺ニ至テ立止リ兩人共左足

ヲ左ノ後へ斜ニ蹈開キ受身ノ

者ハ左足ヲ捕ノ右脇へ蹈込ミ

左手ニテ後ロノ帶ヲ取リ右手ニ

テ前帶ヲ取リ少シ腰ヲ下ル

（一圖參照）捕身ノ者ハ右手ノ

拇指ヲ下ニシテ受身ノ右襟ヲ

横　車

一

Shodan Nagesute First Level Throwing Techniques
横車 *Yoko Guruma* 5/20[26]
Tossing Over the Side of Your Hip Like a Wheel

This technique begins with both combatants standing about 18 feet apart. Both you and the attacker shout and move towards the center, stopping about 3 feet from each other before both step back with your left foot and stand diagonally.

The attacker steps towards your right side and shifts his left foot behind you. He grabs the back of your belt with his left hand and grabs the front of your belt with his right hand. He lowers his hips slightly. (This is shown in the illustration.)

You respond by using the thumb of your right hand to grab the attacker' right collar.

---

[26] The author did not include this technique since he deemed it too dangerous and had too many orally transmitted teachings. However, it was included in the 1893 *book Illustrated Guide to the Inner Mysteries of Tenjin Shinyo School Jujutsu*, by Yoshida Chiharu and Iso Mataemon. These illustrations are from that book.

横一文字ニナルヘシ

（二圖參照）右足ヲ後ヘ開キテ

テ受身ノ左足ヲ外ヨリ拂ヒ

充分ニ右ノ方ヘ崩シテ左足ニ

ノ左ヘ寄ベシ又捕ハ受ノ体ヲ

ヲ崩シ此際受ハ右足ヲ少シ後

体ヲ充分ニ右ヘ向テラ受ノ体

ヘ向左ハ爪先ヲ蹈付踵ヲ揚ゲ

摑ミ右足ノ踵ヲ蹈付爪先ヲ右

二　　車　　横

Shodan Nagesute First Level Throwing Techniques
横車 *Yoko Guruma* 5/20
Tossing Over the Side of Your Hip Like a Wheel

Next, put your weight on the heel of your right foot, and rotate your toes to the right. Put pressure on the toes of your left foot, raising that heel off the ground, and rotate your body to the clockwise. Doing this will break your opponent's balance. This will cause him to move his right foot back and to the left.

Once you have made the attacker shift to the right, sweep his left leg to the outside with your left leg. (This is shown in the illustration.) Finally, step back with your right foot, and return to the stance called Open Side Stance Like the Number One.

片胸捕（カタムナドリ）

○此手モ前ノ如ク双方共ニ声ヲ掛ケ中央

マテ進ミ三尺ヲ巨離テ止リ双方泊（ニ）眼合テ

気ヲ婉ヒ甲ヨリ（ヤー）ト声掛ケル乙方モ（ヤート

答ヘルナリ甲ノ胸ヲ右ノ手ニテ図ノ如ク

取ル甲ハ其手ヲ左リノ手ニテ図ノ如ク下ヨリ

握リ（エイヤ）ト言サマ右拳ヲ上テ乙ノ眉間ヲ

目掛テ打チオロス乙モ（ヤート声掛ケ左リノ

手ニテ受止ノ（但シ指延ス）直ニ其手首ヲ握リ

一寸上へ持テ直又下ケ我ガ左リノ乳ノ所へ引付

我ガ体ヲ左へ捻左ノ足ヲ後へ開クトタシ（マイ）ト言サマ廻シ投ニスルナリ

後ニ直ニ両手ニテ翠ヲ囲ヒ膝ヲ立テ身ヲ構へ甲ノ再ヒ掛カラント心ヲ用ユルナリ

Shodan Nagesute First Level Throwing Techniques
片胸捕 *Katamuna Dori* 6/20
One-Hand on the Chest Seizing Technique

乙 Otsu
You

甲 Koh
Attacker

This technique begins the same way as the previous ones. Both you and the Attacker shout and move towards the center of the training area, stopping about 3 feet apart. You should be looking at each other with Hakugan, white eyes, and be in a state of Nirami-ai, glaring at each other, as you judge each other's fighting spirit.

The attacker shouts *Yaa!* and you respond with *Yaa!* before grabbing the attacker's lapel with your right hand. As the illustration shows, the attacker responds by grabbing that hand from below with his left hand. He then shouts *Eiya!* and raises his right hand, readying to strike down from above to Miken, between your eyebrows. You shout *Yaa!* and block this with your left arm. (Note that your fingers should be extended.) Then immediately grab that wrist.

Lift it up slightly before pulling it down by your left breast. Twist your body to the left as you step back with your left foot. As you do this, shout *Ei!* and throw the Attacker. After throwing, you should have one knee on the ground and one up, with your hands in front of your groin, remaining vigilant in case the Attacker strikes again.

手髮捕 タガミドリ

○此手モ初ハ一ノ一図甲

同シク双方共

二声ヲ掛ケ

中央進

ミ出三尺ヲ

隔テ身ヲ構ヘ

甲ヨリ（ヱイ）ト声掛ケ

右ノ手ニテ乙方ノ手髮ヲ攫ミ（左ノ時ハ左手ハ）

乙方モ之レニ應シ声ヲ掛ケ甲方ノ右肋骨ヘ

我が躰ヲ半身ニ構ヘテ左ノ肘ニテ一本喰ヌル

（形シ）右手ハ畢ヲ圀フニ図ノ如ク殖ニ左ノ手テ×

二　甲

二同シ

乙

×甲ノ右手ヲ取リ巳ガ乳辺ニ引

付（セリ）右ノ足ヲ進ルト二図ニ甲ノ

左ノ肩ニ右ノ手ヲ当ニ図ノ如

クナシ（ヱイヤ言テ甲ノ右ノ腰ノ後ヘ

我右ノ脛ニテ拂フナリ以下片胸

Shodan Nagesute First Level Throwing Techniques
手髪捕 *Tafusa Dori* 7/20
Seized by the Hair Technique

甲**Koh**
Attacker

乙 **Otsu**
You

一
図
甲

This technique starts the same as the previous ones. Both you and the attacker shout and advance towards the center, stopping when you are about 3 feet apart.

The attacker shouts *Ei!* and grabs your hair with his right hand. (He keeps his left hand over his groin.)

You respond by shouting as well and twisting your body so you are standing perpendicular to the attacker. Use your right elbow to drive a hard strike into his right side. (This is *Katachi-nomi*, just showing the movement, not actually striking.)

Your right hand should be guarding your groin.

甲 Koh
Attacker

乙 Otsu
You

図

Then, as shown in the second illustration, you immediate remove the attacker's right hand with your left hand, and pull hit down by your breast. (This means by your left breast.) Step forward with your right foot and strike with your right palm against the attacker's left shoulder. This is shown in the above illustration.

With a shout of *Eiya!* sweep his right thigh back with your right shin. The rest of the technique is like the previous, Kata Muna.

The Illustrations for *Tafusa Dori* from *Illustrated Guide to the Inner Mysteries of Tenjin Shinyo School Jujutsu 1893.*

1	2

小具足 小グソク

○此手モ初同シ　甲

双方同ク壁際ニ

立居申ハ小太

刀ヲ帯シテ掛

声ヲ合ヒ中央ニテ

進ミ三尺ヲ

隔ヱ飜白ニ

合ヒ甲方ヲリ（ヱイヤト言テ小太刀ヲ

以テ乙ノ真向ヲ打ツ（但左ニ

甲ノ打込ム右ノ一腕ノ所ニ右ノ手先ヲ当受止

ルモ受止）右ノ足ヲ豆前ヘ踏出（ヱイヤト言サ

我ガ左ノ手ニテ甲ノ右ノ手首ヲ取テ巻込我ガ右ノ膝ヲ突キ左ヲ

一図

×図ノ如クナシ（エイヤト言テ我ガ左ノ肘

ヲ左ニ捻リ小太刀ノ柄君

ニ挿ク左右ニ開ラクハヅミニ

太刀ハ扱ケ甲ハ後シニ

倒ル、最モ至妙ノ

手ナリ後例ノ如ク

身ヲ構ヘ

甲擧動ヲ

甲ノ背立

又右ノ手ヲ

ミテ甲ノ持

チタル小太

刀ノ柄辺

乙

甲

二

図

Shodan Nagesute First Level Throwing Techniques
小具足 *Kogusoku* 8/20
Fighting (Unarmed Against) Short Weapons

甲 Koh
Attacker

乙 Otsu
You

This technique begins as before, with both you and the attacker standing against opposite walls. The attacker has a Shoto, short sword in his belt. You both shout a Kakegoe and move towards the center of the training area, stopping 3 feet from each other. You both glare at each other before the attacker shouts *Ei!* and, stepping forward, cuts straight down with his short sword. (However, he keeps his left hand covering his groin.)

You respond by shouting *Yaa!* and striking and blocking his upper arm with the end of your right hand. (Your right hand should be in a Shuto, Knife hand, as shown previously in the Shodan technique Tobi Chigai, Jumping Across. Use this to block and stop.)

Next, forward with your right foot one step and, as you shout Eiya! do as shown in the second illustration. Wrap your left arm around the Attacker's right arm and seize his wrist. Then drop down onto your right knee, with your left knee up, behind the Attacker.

乙 **Otsu**
You

甲 **Koh**
Attacker

Then, with your right hand grab the Tsuka, or handle, of his short sword. This is shown in the illustration.

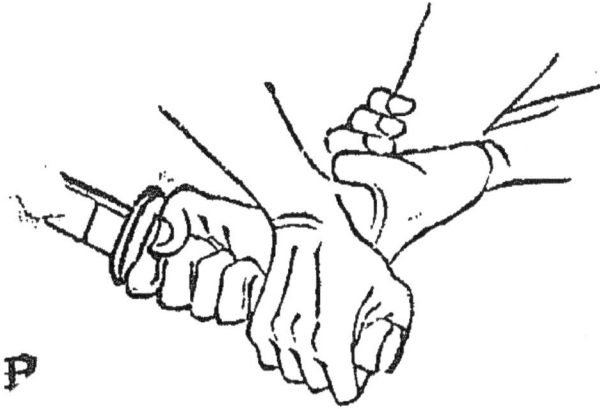

With a shout of *Eiya!* twist your left elbow out to the left while, at the same time, you sweep your right hand out in the opposite direction. By opening up in both directions you yank the sword out of the attacker's hand and throw him backwards. This technique requires a deft timing and movement.

As before you should wait in a ready position until the Attacker gets back to his feet.

The illustrations for *Kogusoku* from *Illustrated Guide to the Inner Mysteries of Tenjin Shinyo School Jujutsu 1893*.

1	2

三十七

腰苅捨 コシカリステ

○此手モ前ハ同シ双方声掛ケ中央マテ

進寄リ三尺巨離テ止ル（詞ヲ見）甲方

（ヲイ）ト言テ右ノ足ヲ一足前ニ進ミ右ノ手

ニテ乙方ノ右ノ手先ヲ握リ我カ前ヘ

腰ニ付テ引キ左手ヲ乙ノ左ノ肩ニ掛

ケ乙ヲ撥ネトス図ノ如ク乙方ハ右ノ手

ヲ引ル、時ニ右ノ足ヲ「足甲ノ後ヘ踏

出シ前ノ初段ノ連柏子ノ如ク右ノ手

取ラレタル侭（ヲイ）ト言サマ躰ヲ半身ニ構（ヲ圖ヒ）中腰ニナリ

（ヤ）言テ後ヘ手ヲ押シテ投出スナリ

乙

甲

腰苅捨 *Koshi Kari Sute* 9/20
Hip Sweep and Dispose

甲**Koh**
Attacker

乙

乙 **Otsu**
You

This technique begins the same way as the previous ones. Both you and the attacker shout before advancing towards the center, stopping about 3 feet apart. (You and the attacker should be glaring at each other.)

The attacker steps forward with his right foot and shouts *Ei!* and seizes the end of your right hand with his right hand. He then moves in front of you and tries to throw you by setting his hip against you while taking hold of your left shoulder with his left arm and pulling. This is shown in the illustration.

When the attacker pulls your right arm, step further behind him with your right foot. Then, as was shown in the Renhysoshi technique in Initial Stage, drop your hips down into Chugoshi, as if you are sitting down. With the attacker still holding your arm, twist your body to the side, (while keeping your left hand covering your groin) and shout Yaa! as you shove the attacker down behind you.

獨鈷（トッコ）

○此手モ双方声ヲ掛ケ中央マテ進ミ双方

行違ヒ（初段ノヨリ投ノ）甲方ヨリ（ヤ）ト声掛ケ
スレチカ（如ク行違ッテ）
ル乙方ハ甲ノ後ニ 廻リ（ヤ）ト言ヒサマ新城ニ

両手 平掌ニテ甲ノ両耳ヲ縦ト打チ
（好ハ）直ニ左ノ手先ヲ甲ノ頤ニ掛ケ其肘
ヲ甲ノ肩ヨリ背ノ紋所ノ辺ヘカケテ付ケ

右ノ手ヲ以テ甲ノ頭ヲ持チ一寸ト起スト見セ首ヲ
左ヘ捻ル（甲ノ体ヲ崩スタメ）（エイヤ）ト言ヒサマ我右ノ足ヲ右後ヘ開キ
トタニ我前ハ甲ヲ倒スナリ以後前ニ同シ

小手返

引落シ

此両手ハ危険ニシテ手モ亦
入クミ草頭画伏ニ尽シカタシ
略之

Shodan Nagesute First Level Throwing Techniques
獨鈷 *Dokko* 10/20
Vajra[27]

This technique begins as the previous ones do, with both combatants shouting and then moving towards the center of the training area and walking past each other (This attack that occurs while crossing paths is similar to Eri Nage, Collar Throw, from the Initial Stage.)

The attacker shouts *Ya!* and you slip around behind him and shout *Yaa!* Then suddenly slap your palms over his ears. (When training you stop about one inch short of actually hitting.) Then immediately grab his jaw with your left hand, resting that arm on his shoulder so your elbow is at the Mondokoro, or house seal on the back of his shirt.

---

[27] A vajra is a ritual weapon with the indestructibility of a diamond the irresistible power of a thunderbolt. It is a Kyusho, vital point in a chart at the end of this book.

Use your right hand to hold the attacker's head, raising and twisting it slightly to the left. (This will break his balance.)

Illustration for *Dokko* from *Illustrated Guide to the Inner Mysteries of Tenjin Shinyo School Jujutsu 1893*.

Finally, shout *Eiya!* as you step back and to the right with your right foot, dropping the attacker in front of you. The rest of the technique proceeds as the previous ones do.

As was mentioned in the <u>Table Of Contents</u>, the techniques Kotegaeshi, Small Hand Reverse, and Shikiotoshi, Pull Down, are not included in this book. The reason is these techniques are both dangerous and contains many oral traditions. As they are beyond the ability of the artist to capture, they have been abbreviated.

Translator's Note:
While the description does not mention it specifically, the vital point Dokko, the hollow behind the ear, is where your left hand should be placed.

Dokko

一小手返　コテガヘシ

此手合ノ形ハ三間計リ離レテ

直立ナシ互ニ聲ヲ掛テ進ミ請

身ノ者カ我前へ捕身ノ者ガ近

付タル際右足ヲ少シ前へ進テ

捕身ノ左右ノ手首ヲ左右ノ手

ニテ摑ムナリ（圖参照）此時捕

身ノ方ハ左足ヲ少シ斜ニ開キ

乍ラ右手ノ指先ヲ我左リ手ノ

肩ノ邊へ向ルト同時ニ臂ヲ請

身ノ顔ノ方へ張レバ自然ト手

小

手

返

一

Shodan Nagesute First Level Throwing Techniques
小手返 *Kotegaeshi* 11/20[28]
Reversing the Wrists

Both you and the attacker start about 3 Kan, 18 feet/5.4 meters apart. You shout at each other and then move towards the center of the training area. As you begin to move towards the attacker, he steps forward slightly with his right foot and seizes both your wrists. (This is shown in the illustration.)

You respond by stepping diagonally back and left with your left foot, extending the fingers of your right hand and thrusting your elbow towards the attacker's face pointing while whipping your fingers up to your left shoulder. This will naturally free your hand.

---

[28] Kotegaeshi, Small Hand Reverse, and Shikiotoshi, Pull Down were not included in *Bujutsu* since they were deemed too dangerous and difficult to describe. The following illustrations and explanation are from *Illustrated Guide to the Inner Mysteries of Tenjin Shinyo School Jujutsu 1893.*

ガ放レル故直ニ指ヲ伸シタル

マ、掌ノ甲端ノ方ニテ烏兎ヲ

當ルト同時ニ左リノ臂ヨリ指小

先迄眞直ニ掌ヲ我方ヘ向テ上

（一圖参照）右手ニテ左リノ手手返

首ニ掛居ル受ノ右手ノ掌ヲ持

次ニ左手モ右手ト同樣ニ掌ヘ返

掛其マ、左リノ後ヘ充分ニ廻

リ請身ノ左リ脇ヘ並ブ樣ニ踏二

開キテ請身ノ掌ノ左リヘ逆ニ

返シテ投直ニ手ヲ放スナリ

After that, while keeping your fingers extended, strike to Uto, between the eyebrows, with the bottom of your right palm.[29] At the same time, straighten your left arm from the elbow to the tips of your fingers and thrust it up and towards yourself. (This is shown in the second illustration.) With your right hand, seize the back of the attacker's right hand, which is holding your left wrist.

Next, use your left hand to seize the back of the attacker's left hand. Rotate your body counterclockwise sufficiently, so you are aligned with the attacker's left side. Step out to the side while holding both the attacker's hands. Finally, rotate counterclockwise and throw your opponent by rotating the back of his hands so they are reversed.

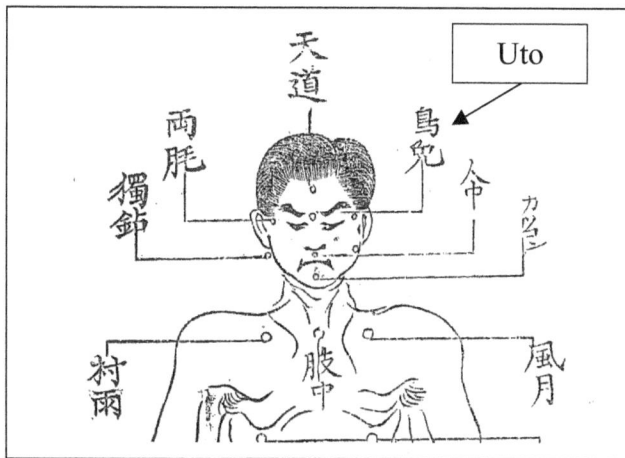

---

[29] The striking point Uto 烏兎 "crow and rabbit" is the center point of the face. It is between the eyes at the top of the nose. Striking this point will blind an opponent and cause them to lose consciousness. Even a glancing blow will cause a bloody nose.

-Jujutsu Textbook (1912)

一引　落　シキヲトシ

此手合ノ形ハ三間計リ離レテ

直立ナシ互ニ聲ヲ懸テ進寄請

身ノ者近附タル時左右ノ手ニ

テ捕身ノ左右ノ襟ヲ一ニナシ

右手ニテ絞ニ取リ左足ヲ斜メ

ニ開キナガラ右手ニテ少シヲ

スヘシ此際捕身ノ方ハ左手ニ

テ受身ノ右ノ手首ヲ下ヨリシ

カト掴ミ左足ヲ左リノ後へ斜

メニ三尺計リ蹈開キナガラ右

手ヲ延揃ヘ甲端ノ方ニテ受身

ノ烏兎ヲ當テ（二圖參照）直ニ

引　落　一

Shodan Nagesute First Level Throwing Techniques
引落 *Shiki Otoshi* 12/20
Pulling Down

This technique begins with you and the attacker standing about 18 feet apart. You both shout and begin advancing towards the center until the attacker reaches out and seizes your collar with both hands. His hand grip parallel to each other. Next, he uses his right hand to choke you by stepping diagonally with his left foot and pushing with his right hand.

You respond by grabbing his right wrist from below with your left hand. Step 2~3 feet diagonally backwards and to the left with your left foot. As you do this extend your right hand with the fingers together and strike to Uto, the spot between the eyebrows, with the base of your palm. (This is shown in the illustration.)

甲端ヲ右ノ尺澤ヘ掛テ少シ腰

ヲ下ナガラ尺澤ヲシカト押付

此時受身ノ者ハ右膝ヲ前ヘツ

クヘシ（二圖参照）捕身ノ方ハ

其マ、右足ヲ左足ノ踵ノ際ヘ

ヨセテ次ニ左足ヲ左リ方ヘ斜

メニ蹈開キナガラ左手ニテ我

襟ヲ持右手ノ甲端ニテ尺澤ヲ

押テ落スヘシ此時受身ノ者ハ

手ヲ引ル、ニ順テ仰向ニ倒レ

ル者トス

引　落　二

Shodan Nagesute First Level Throwing Techniques
引落 *Shiki Otoshi* 12/20
Pulling Down

Then immediately place your palm on the attacker's Shaku Taku, a vital point on the inside of the elbow. Drop your hips slightly as you shove down on Shaku Taku. This will force your opponent down onto his right knee. (This is shown in the second illustration.) You then shift your right foot over so it is directly beside your left heel, then step out diagonally left with your left foot.

As you do this grab the back of the attacker's right hand, which is holding your lapel with your left hand and push down on Shaku Taku with your right to drop the attacker. As you pull his arm he should turn so he falls face up.

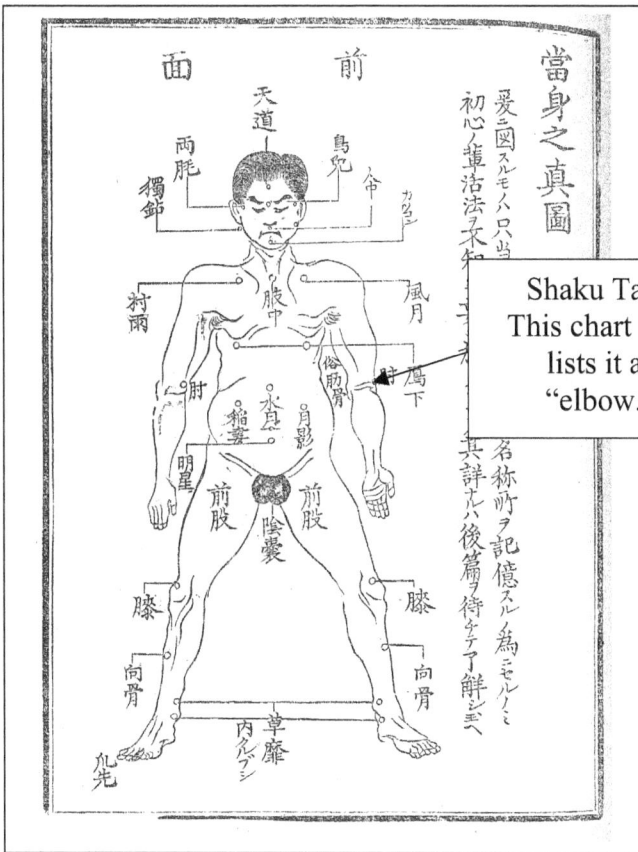

Shaku Taku
This chart only lists it as "elbow."

手操 タグリ

○此手モ前ハ同シ例ノ如ク双方共ニ

声ヲ掛ケ中央マテ進寄ニ三尺ヲ距レテ

双方互ニ顔ヲ見合居ル甲方ヨリ（エイト

言テ乙方ノ胸ヲ左リ手ニテ取ル（グミ之

（右手ハ遅／右図ヲ見ヨ）乙方モ（ヤレ）ト答ヘテ我ガ

棒ヲ半身ニ搆（サマ直ニ右ノ足ヲ甲ノ

後ニ踏出シテ其場ニ右ノ膝ヲ突キ左リ

戻ヲ立膝ニナシ図ノ如クシ（ヱイ）ト言サマ甲ノ囲ニ

タル縄（逆）ヲ左ノ手ノ内ヨリ我ガ右手指先ヲ延ニテ

上リ打下シ図ノ如クナシ甲ノ首（頸）ヲ掴ミ左ノ手ガ甲ノ囲ニ

ノ股ヲ啜上（ヱヤ）ノ声ト共ニ我ガ右横後ヘ投倒ナリ

甲

乙

Shodan Nagesute First Level Throwing Techniques
手繰 *Taguri* 13/20
Controlling with the Hands

Left: Illustration One   Right: 1893 version

This technique begins as the previous ones do, with both you and the attacker shouting before advancing towards the center and stopping 3 feet apart. You and the attacker look each other in the face. He shouts *Ei!* and grabs your lapel with this left hand. (Specifically, he grabs your *Munagura*, lapel.) (Your opponent keeps his right hand over his groin.)

You respond by turning your body perpendicular to the attacker while shouting *Yaa!* Then step behind him with your right foot. While keeping your left knee upright drop down onto your right knee. This is shown in the illustration.

With a shout of Eiya! thread your left hand inside the left arm of the Attacker, that is gripping your chest.(Your Munagura, lapel.) Extend the fingers of your right hand and use your arm to strike downward onto the Attacker's right arm as shown in the illustration. Then grab his back (The Mondokoro, family crest.) while your left hand slips under his left thigh. Scoop his leg up and throw him back and to your right side with a shout of Eiya!

捨身 ステミ

○此手ハ前ノ初段ノ立合ノ絹漉ノ如クナリ甲方ヨリ

（エヽト声ヲ掛ケ壁際ヨリ三四尺前ニ出ル乙方モ掛声ニ應シ（ヤー）ト信テ

小走ニテツカヽ（ト甲方前ニ二間許リノ所ニ至リ止ル

甲ヨリ（ヤ）ト声ヲ掛乙モ同シク答ヘ右手ニ

甲ノ股ヘ差入図ノ如クナシ（此時甲ハ

取リ我ヵ左リノ膝ヲ突キ右ノ足ヲ

甲ノ胸グラヲ取リ左ノ手ニテ前帯ヲ

右ノ拳ヲ搦シ乙ノ頭腦ヲ打ントス乙モト

言サマ右ノ手ヲ押シ左ノ前ヘ引躰返シ後ヘ返ルヽ此トタニ

右足先ニテ甲ノ左内股ヲスヽ上ゲ向（速々投ルヽイナ直ニ起直リテ甲ノ方ニ向テ左膝ヲ

突キ右ヲ立両手ニテ畢ヲ囲ヒ身構ヲナス

甲

乙

Shodan Nagesute First Level Throwing Techniques
捨身 *Sutemi* 14/20
Sacrifice Throw

This technique is quite similar to Kinu Moguri from the First Stage Standing Techniques. The Attacker shouts Eiya! and advances 3 or 4 Shaku, 90 ~ 120 centimeters away from the wall. You respond with a shout of Yaa! and do a short dash forward, with your feet making a Zuku Zuku sound of feet moving quickly. Stop about 2 Shaku, 60 centimeters in front of the Attacker. The Attacker shouts Yaa! and you respond in the same fashion, before grabbing his collar with your right hand. Your left hand takes hold of the front of his belt.

Plant your left knee on the round and extend your right leg between the Attacker's legs. (At this point the Attacker has raised his right fist and is preparing to strike you in the head.) Shout Eiya! as you push with your right hand and pull with your left, bending your body back as you do so. As you roll back, use the end of your right leg to lift the inside of the Attacker's left thigh, which will cause him to be thrown far behind you. Then immediately rise and face the Attacker with your left knee on the ground and your right knee upright. Both hands should be covering the groin.

(When being thrown, the Attacker should try and keep his chin tucked in.)

Kudari Fuji

This next technique is dangerous so it has been abbreviated.[30]

---

[30] This technique is missing.

腕縅 ウデカラミ、

○此手モ双方共ニ前ハ同シ構ナリ互ニ声ヲ掛ケ

中央ニテ進ミ三尺許ノ巨離ヲ計リテ

止リ甲方ヨリ（アイダ）ト声掛ケ右ノ拳ヲ

振上ケ左手ハ罾ヲ曲ヒ乙方ノ脳部

ヲ見込打掛ル乙方モ（ヤ）ト答へ

直チニ右ノ足ヲ甲ノ右足ノ横へ踏出シ

其打礎シクル甲ノ手首ヲ左ノ手ヲ夾

箸ニナシテ受止メスカサズ其手ヲ掴ミ（ヤ）ト言ヒ

我カ右手ヲ甲ノ右脇ノ下ヨリ差入レ甲ノ右掌へ右ノ指ヲ掛縅タマヽ我カト

共ニ図ノ如ク向へ押捻ルナリ（押付ルベカラス）甲ハ早ク何ナ共打テ頁ヲ報

下り藤 *Kudari Fuji* 15/20
Descending Wisteria (Not Included)[31]

腕緘 *Ude Karami* 16/20
Arm Tangle

This technique begins the same way as the previous ones, with both you and the attacker shouting and moving towards the center of the training area, stopping about 3 feet from each other. The attacker shouts *Ei!* and raises his right fist to strike, keeping his left hand covering his groin. He is aiming to strike you in the head. You respond with a shout of *Yaa!* and step forward with your right foot, so you are beside the attacker's right foot.

Stop the attacker's arm with your left hand in Yahazu, or shaped like the nock of an arrow, then immediately seize that wrist. With a shout of *Yaa!* slip your right arm under the attacker's right arm and grab his palm, gripping it with your fingers. Having entangled his arm use your power to twist and push him backwards. This is shown in the illustration. (You should be careful not to push too hard when doing this technique.)

The attacker should immediately strike the ground to show that he has been defeated.

---

[31] Probably in the reprint…

矢筈 ヤハツ

○此手モ最初同シ双方声ヲ掛ケ罩ヲ囲ヒ乍ラ中央ニ

進出テ三尺許ノ巨離ニ至リ（此手ハ乙方ヨリ扚掛ルナリ）

甲方ヨリ（ヤー）ト声ヲ掛ケル乙方ハ（エイ）

ト答ヘ直ニ甲方ノ咽喉ヘ右ノ手ヲ矢

筈ニシテ押当左ノ手ニテ甲ノ手ヲ

首ヲ取リ我カ左ノ乳ノ所ヘ引付ケ

引（押ト引トナリ）図ノ如クヌ甲方ハ我カ左ノ掌

三ッ乙ガ咽ヲ押ヘタル手先ヲ横ニ（右）打拂フ

乙ハ其ヲ拂ヒタル手先ヲ甲ノ右ノ肩口ニ当（エイ）言フサマ

右ノ腕ニテ甲ノ右ノ股後ヲ拂ヒ（エイ）ト投ルナリ此時直ニ身ヲ左ヘ開キ

両手ニ罩ヲ囲ヒ身構ヲナス

両手捕
此手ハ余程込入リテ文字
画図ニテ解シ難シ故省ク

Shodan Nagesute First Level Throwing Techniques
矢筈 *Yahazu* 17/20
Nock of an Arrow

This technique also begins with both you and the attacker shouting and moving towards the center of the training area. As you move forward your hands should be covering your groin. You both stop when you are about 3 feet apart. (In this technique you are striking first.)

The attacker shouts *Yaa!* and you respond with a shout of *Eiya!* and immediately shove the attacker in the throat with your right hand in Yahazu, shaped like the nock of an arrow. Next, seize the attacker's right wrist with your left hand and yank it up beside your left breast. (This means you are both pulling and pushing the attacker.) This is shown in the illustration.

Yahazu from *Illustrated Guide to the Inner Mysteries of Tenjin Shinyo School Jujutsu 1893.*

In response, the attacker uses his left palm to knock away your hand on his throat. This is an *Uchi harau*, strike and sweep using the palm of his left hand. (He sweeps to the right.) When your hand is swept aside, use it to grab the attacker's right shoulder. Shout *Ei!* and sweep the back of his right thigh with your right leg. Throw him down with a shout of *Eiya!* The moment you complete this action you should rotate your body counterclockwise, and take a stance with your hands covering your groin.

Ryote Dori, Two Handed Capture.

This technique comprises a great many steps, which means it is difficult to capture with art. For that reason, it has not been included.

一兩手捕 リョウテドリ

此手合ノ形ハ中三間計リ隔テ
直立ナシ捕身ノ方ハ受身ノ前
迄進ミ行受身ノ者ハ少シク右
足ヲ開キテ捕身ノ左右ノ手首
ヲシカト握ルヘシ此際捕身ノ
方ハ左右ノ指先ヲ延シテ腰ヲ
下ケ乍ラ地ヘ附位ニ体ノ力ニ
テ下ヘ突キ左ノ膝頭ニテ受身
ノ左手ノ尺澤ヲ押シテ取リ
（一圖参照）直ニ左ヘ廻テ受身
ヲ背後ニノ左膝ヲ突キ乍ラ左
手ノ指ヲ延タルナリ内平ヲ上

両　手　捕　一

236

Shodan Nagesute First Level Throwing Techniques
両手捕 *Ryote Dori* 18/20
Two-Handed Capture

This technique begins with you and the attacker standing about 18 feet apart. As you advance towards the attacker he takes a short step out to the right and suddenly seizing your wrists.

You respond by extending the fingers of both hands and dropping your hips. Use your weight to force the attacker down towards the ground. Plant your left kneecap on the point *Shaku Taku,* forearm bone, on the attacker's left arm to break his grip. (This is shown in the first illustration.) Then immediately rotate clockwise so your back is against the attacker and plant your right knee on the ground.

両手捕二

二向テ手首ヲ握リ右手ニテ受
身ノ右ノ肩先ヲ摑横一文字ニ
蹈開キ（二圖参照）腰ヲ下タル
マ、受ノ右腕ヲ前ヘ引投テ直
ニ腕ヲ我胸部ノ方ヘ引揚ケル
ナリ此際受身ノ者ハ投ラレテ
起上ルニ非ス左手ノ平ニテ下
ヲ打テ止ルベシ此形ノ投ニ下
ヲ放ス人モ在レモ流祖ノ掟ニ
依ル時ハ投テ直ニ手前ノ方ヘ
引上ケルヲ宜シトス

As you do this, extend the fingers of your left hand and take hold of the attacker's right wrist, which is holding your lapel. Reach up and grab his right shoulder with your right hand. Shift your left leg out to the side so you are in Yoko Ichimonji, making a line like the Kanji for one一. (This is shown in the second illustration.) Keeping your hips low, pull the attacker's right arm forward, throwing him. Then immediately pull his arm towards your chest, lifting him up.

Since the attacker has been thrown, he is unable to rise so he should strike the ground with the palm of his hand to show he is defeated.

When doing this Kata there are some people that release the arm, however the founder of the school dictated that you should pull the arm up after throwing, so you should do so.

両柄捕 リャウツカドリ

此手モ最初ノ
ハ同シク掛声ニツレ
中央マテ進ミ
立(二図ノ如シ)(双方共)
顔見合ヒ氣ヲ付ケ
甲刀(子)ト言マ乙方
人太刀ヲ柄ヲ両手ニテ握リ引拔クトスル乙方ハ(ヤット答テ
左手親指ヲ鍔ニ掛ケ抓迚ヘテ抜カセマジトナス右手
ニテ甲ノ面(眼ザシヲナス(右角ヲ死)ヲ跳ネ
(但形)一図ノ如シ 直ニ我右手ニテ甲ノ持ナタル柄頭ヲ握リ
其後右ノ後ノ方(ヘ)(子ヤト信サマ右手足ヲ同ニ開ク

一図
甲 乙 一

二図ノ如クシ甲ノ左ノ肩ヘ左リノ
手ヲ掛ケ左足ヲ甲ノ右ノ股後ニ立
中腰ニテレ(子ヤト言マ甲ヲ我左
リ後ヘ捻投ニ倒スナリ
此時柄頭ヲ握リタル佐甲ニ鍔リ
贖ヲヌヤウ上ヘ持クルナリ

(此上ハ甲ハ我ガ)(体ニ近ツクシ)我体ヲ半身ニ構ヘ

二図
甲 乙

Shodan Nagesute First Level Throwing Techniques
両柄捕 *Ryotsuka Dori* 19/20
Two-Handed Sword Handle Capture

This technique begins the same. Both combatants shout a Kakegoe and move towards the center of the training area. (Stopping 3 feet apart.) (You are wearing an *O-Dachi*, longsword.) You and the attacker lock eyes, each judging each other's martial readiness. The attacker shouts *Ei!* and seizes the handle of your sword with both hands, and tries to yank it free.

Illustration 1

You respond by shouting *Yaa!* and hooking your thumb over the *Tsuba*, hand guard, preventing your sword from being taken. Use your right hand to do a *Metsubushi*, feint at the eyes. (This will cause your attacker to turn his face to the right.)

Next, kick your Attacker in the groin with your right foot. (This is *Katachi Nomi*, or just showing the movement and not actually striking.) This is shown in the first illustration.

Illustration 2

Then immediately move your right hand down to grab the *Tsuka-gashira*, pommel of your sword that the attacker is holding. Next, while shouting *Eiya!* take a (big) step backwards and to your right with your right foot, moving your right arm in unison. (Ensure that you remain close to the attacker's body.) This will turn your body perpendicular to the attacker's.

Next, as shown in the second illustration, seize the attacker's left shoulder with your left hand and plant your left leg behind his right thigh. Drop your hips low as if you are going to sit down, and with a shout of *Eiya!* twist and throw the attacker behind you and to the left. You should maintain your grip on the *Tsuka-gashira* the whole time, preventing the *Kojiri*, the metal cap on the pommel, from striking the attacker.

The Illustration for *Ryote Dori* from *Illustrated Guide to the Inner Mysteries of Tenjin Shinyo School Jujutsu 1893.*

○後捕 ウシロドリ

○此手ハ乙方ガ中央ニ両手ニテ我ガ

澤ヲ囲ヒ絎ノ向キテ立居ル甲方ヨリ（ヱイ）

言サマ乙方ノ諿ヨリ抱込ナリ乙ハ（ヱ王）

答ヘテ其鶙附レタル侭右ノ足ヲ

甲ノ左リノ足ノウシロヘ廻シ両肘ヲ

張リ躰ヲ一文字ニ構ヲナシ右ノ手

ヲ抜出シ図ノ如クナシテ前ノ横

車ノ如ク甲ノ胸ヘ我カ右ノ肘ヲ上ヨリ

打落スナリ以下横車ニ同シ

投捨ノ部終

Shodan Nagesute First Level Throwing Techniques
後捕 *Ushiro Dori* 20/20
Seized From Behind

This technique being with you standing in the middle of the training area, facing straight ahead with both hands covering your groin. The Attacker shouts Ei! and, moving up from behind, wraps his arms around you.

You respond with a shout of Eiya! and, with the Attacker still holding your arms, slip your right leg behind the Attacker's left leg. Shove both elbows out, making your body a straight like the Kanji for one一. Pull your right arm free as shown in the illustration. The technique finishes like Yoko Guruma, Tossing Over the Side of Your Hip Like a Wheel. You slam your right elbow down into his chest and finish as described in Koshi Guruma.

都て乱撲ハ
此ノ一段ニテ
浪練ニテ
前ニ示ス
師許ヲ得テ
可捕ベシ
双方共ニ
足ノ動キ肝要ノ手ナリ

乱撲十二本ノ図

乳取ハ双方トモニ
其左右乳ノ
辺ヲ君手ニテ
掴ミ左キ
ハ右ノヤツテ或ハ
袖下ヲ取ル
双方共ニ
足ノ動キ肝要ノ手ナリ

襟捕此図ノ如ク
右ノ手ニテ咽
ノ辺ヲ襟ヲ
取リヤツテ
或ハ袖ヲ掴ミ
後ノ袖ヲ掴ミ
足ニテ向ノ
足ヲ挑ヒ
腰投ケ専ノ
手ナリ
左ミテモ同シ理ナリ

小手シギ此ハ片方ニテ
胸グラヲ取リ又一方
ハ我ガ躰ヲ
少コミ
其胸グラノ
手ヲ両手
ニテ抱メテ
図ノ如クナスナリ
此手逆手ナル故ニ不熟ノ者ハ其危険ヘ

突込ミ
此手ハ片
テ両テヲ
咽辺ヲ取リ
図ノ如ク片方
ノ顔ヲ向ル
方ヘヨリ
突込ミ片方ニテ又
同シ此手ヲ掛ルトキハ
我襟ヲハ持振挑ヒテメサセヌヤウニスベシ

引キヲルノ

Randori Free Sparring
12 Illustrated Techniques

Randori, free sparring, is not limited to the following techniques. You should thoroughly train the techniques introduced in the previous sections and, after getting permission from your instructor, train them with a partner. Depending on your level of skill, you should be able to develop a multitude of variations. This will help you develop until you are ready to try a real duel. Therefore, the following techniques should be considered starting points for free sparring.

Nyu Tori
Chest Grab

This is grabbing with both hands. Seize the left side of your opponent's chest with your right hand. With your left hand, grab his *Yatsuguchi*, the opening in the shirt under the armpit, or the end of his right sleeve.[32] It is essential that both combatants move their feet effectively.

---

[32] Early Jujutsu shirts had an open slit under the armpit, these were later eliminated.

Eri Tori
Seizing the Collar

As the illustration shows, use your right hand to grab both sides of your opponent's collar just under his throat. With the other hand grab *Yatsuguchi*, the slit underneath the armpit, or the end of the sleeve. Then step behind your opponent and sweep him to the ground. This technique uses *Koshi Nage*, Hip Throw. The technique can be done the same way on the opponent's left side.

Kote Shigi
Lower Arm Twist

In this technique your opponent attacks by seizing both lapels under your throat and attempts to wrap his other arm around your body. Scoop the elbow of the arm holding your collar up from below with both hands. This is shown in the illustration.

Since this technique uses a joint lock, it is extremely dangerous for inexperienced learners.

Tsuki Komi
Piercing In

Seize your opponent's lapels with both hands near his throat. With one hand push forward, thereby forcing his face away while pulling with the other hand. While you are doing this ensure that your opponent, who is gripping your collar below your throat, does not sweep your hands way and choke you.

胴ノ此手ハ組合

テ下ニナリタル時我カ
足上ニナリタル者ハ
腰ヲ　　　　図ノ
如クナシテメルヘ
又上ノ者ハ我体
ヲ半身ニナシ下
腹ニカヲスメテモ
キカヌヤウニスルナリ

捨身

此手ハ初
段立合ニアリ又
投捨ノ部ニモアリ
別ニ詳記セス只相手ノ
スキヲ見ルヲ肝要ノ手也

突込此手モ組合
上ノ者ハ前ニ突込
ノ手ナス下ノ者亦
組ニ入レタリトモ突
込ス或ハ襟ノ等掛
ラル、ナリ柔術ハ
都テ下ニナリタル者
ノ伎ニ妙アリ
ワサ

腕シギ組合テ一方
胸グラヲ取リ其後
握リ躰ヲ向ニナシ
両足ヲ図ノ如ク相手
ノ咽ノ辺ヘ押当我カ
体ヲ延シ十分ニ引タリ
足ハ成文對手首ノ方ヘ
囲スヘシ尤逆手ナレバ注意スヘシ

直ニ両手ニテ手首ヲ
ニタヲレヒト有トキ

Dojime
Waist Strangle

Use this technique when you end up on the bottom during a match. Force your legs up and wrap them around your opponent's waist as shown in the illustration.

If you are on top in this situation, then you should twist your body to the side and put power in your lower abdomen to render the leg strangle ineffective.

Sutemi
Sacrifice Throw

This technique was covered in both the First Level Standing Techniques as well as Throwing Techniques, therefore the details will not be repeated. However, watching your opponent carefully for an opening is the most important point in this technique.

Tsuki Komi
Piercing In

This is also a technique you use when on the ground. Your opponent used the previously introduced Piercing In technique to topple you. Though you are on the bottom you are continuing to fight. You attempt to use Piercing In from the ground or *Eri Jime*, Collar Choke, or any other technique.

In Jujutsu there are many techniques available to the person on the bottom.

Ude Shigi
Arm Lock

You began with on hand on the opponent's collar. Seeing a chance, you grab his wrist with both hands and roll onto your back. Position your legs as shown, with one pressing into his throat. Then stretch your body as you pull his arm. Your legs should cover your opponent's throat. Since you are bending his elbow joint back, use caution.

三十八

肌我捕　此手乳捕双方組合ノ
時三方向ヘギ成ルコアリ直ニ左リ
腕ヲ咽ニ掛ケテ図ノ如クシ
背ニ裏リ掛リ右ノ肘ヲ左リ
手先ニテ持右ノ手先
ニテ下ニナシタル者ハ
頭ノ後ヨリ押付ルヽナリ
此如何ニシテモ外サレヌ
故ハタカニ糸レバ直ニ手ヲ打ナリ

捨身投構　左右共ニ同シ
我ヨリ先ヘ
倒レテ敵ヲ
投ルナリ
双方組合テ
居ル内ニ敵ノ
足蹴ヲ我ガ足ノ
蹠ニテ踏堪テ我ヨリ
先ヘタフレテ敵ヲ投倒スナリ

腰投　双方組合ノ節ニ都合
ニ因リ双方ヨリ
腰投ヲ掛ル
ヲ専ラ下ニナスコ
アリ
前ノ初段ノ部
投捨ノ手ニテ腰
カニ入テ投ルナリ最早業
ヲ要ス

背負投
此手ハ組合フ都
合ニ依リテ對手
ノ手首ヲ取
リ図ノ如ク
肩ニ掛ケテ
向ヘ投出ス
ナリ尤危険
ナル手故不熟ノ者ハ

×ハ此手ハ省クベシ

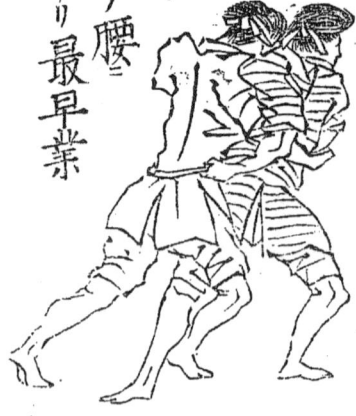

Hadaga Tori
Naked Self-Choke

This technique begins with both combatants using *Nyu Tori*, Chest Grab. You twist around behind your opponent and immediately wrap your left arm around his throat. This is shown in the illustration.

As you lean forward against his back, reach over and grab your right elbow with your left hand. Your right forearm should push against the back of his neck. Once you have this lock, your opponent cannot escape. Thus, even if you were naked you could attack with this technique.

Sutemi Nage Kamae
Preparing for Sacrifice Throw Stance

Both you and your opponent are planning on attacking with a Sacrifice Throw. Each wants to be the first to throw down their opponent. In the middle of your struggle, your opponent tries to sweep your leg using the sole of his foot. You are able to withstand this attack and throw your opponent to the ground.

Koshi Nage
Hip Throw

This starts with both you and your opponent struggling to attack. In this situation you have both decided to rely only on *Koshi Nage* to win.

As was shown in the First Level Throwing Techniques, you put power in your hips and throw. It is essential that you do this as fast as possible.

Seoi Nage
Shoulder Throw

This technique also starts out from *Kumiai*, with both combatants holding each other. When you see an opening, seize your opponent's wrist and load his arm on your shoulder as shown in the illustration. Then throw him down in front of you.

Note that this technique is very dangerous, so inexperienced learners should not do it.

Sword

○撃劍ノ部

凡劍法ノ大用ハ世人知ル處其概畧ハ卷首ニ記セリ然而

其詳カナルコ後ニ解クベシ只爰ニ載ルモノ其兩三ノミ

○短刀ヲ学フヘキ辨

短刀ハ長劔ト咬ヘテ常ニ練習スベシ嘗テ先師長短仁一味

ト説レタリ長劍ト勝頁ヲ試ハルニ短ニシテ大ニ利アリ

濱際ニ於テモ家内ノ闘手抔ハ長劍ニ窄アッテ短刀ニ渕

アリ此ニ玄フ短刀ハ乃チ脇差ナリ竹刀ハ一尺八寸以下

ニ造ルベシ丸モ鍔ヲ付ルナリ用法ハ後圖ノ如ク右手ニ

柄ヲ握リ左手ハ鉾ニ添テ腰下ニアリ右足ヲ進メ輕ク踏

ミ左足ハ其趾部ノミヲ踏ミ兩足地ニ据附カズ偏ヨラス

シテ相手ヘ動ニ臨フナリ對手術ヲ施スニ應シ身ヲ捨テ

How to Duel in Gekken, Japanese Fencing[33]

Most people today understand the main purpose of Kenpo, sword fighting. While the underlying philosophy was described at the beginning of this book, the following section will give detailed examples, however there will only be two or three techniques.

You Should Learn How to Fight With a Tanto, Short Sword

Training with the *Tanto*, Short Sword, should be included as part of your training curriculum with the *Choto*, Long Sword. Recall the previously introduced words, *The teacher who describes the benefits and cautions regarding using a longer blade and a shorter blade.*[34] Simply put, if your opponent seeks to duel with a long-bladed sword, using a shorter blade will give you the advantage. In reality, if there is a duel inside your house then a long sword will be a hindrance, thus a Tanto, short sword would be the most practical in that situation.

When speaking of the *Tanto*, Short Sword, what I am actually referring to is the Wakizashi, the second, shorter sword worn by Samurai. When making a bamboo training short sword, it should be shorter than 1 Shaku 8 Sun, or 21 inches/54 centimeters. After making it, attach a *Tsuba*, hand guard where the handle meets the blade.

The way this weapon is used is as shown in illustrations in the following section, however I will describe it here: hold the handle in your right hand with your left hand against your lower hip. Advance with your right foot, placing it lightly on the ground with your left foot following and stepping in the footprints of where you stepped with your right foot. Both feet should never be on the ground at the same time. You should be leaning towards your opponent, watching for signs of how he is planning to move or attack.

---

[33]Following the Meiji Restoration sword instructors found themselves with no source of income. Carrying swords was banned and the new military and police were adopting Western-style dress and weaponry. Several former sword instructors created a sword dueling tournament based on Sumo events. The first ten-day event, held in 1873, completely sold out.

[34] This is not referring to a short sword versus a long sword, rather a long sword with a longer blade versus one with a shorter blade.

手元ニ飛込ミ勝ヲ取ルナリ對手上段ニ構ヘテ打込ン

ト七ハ吾ハ下段ニ構ヘ打込ム竹刀ヲ攦ジナガラ手元ヘ飛

込ミ左手ヲ以テ對手ノ小手或ハ竹刀ノ柄ヲ取リ首抽ノ

働キヲ妨ケナカラ面部ヲ突クベシ又對手下段ニ構ヘハ

吾ハ揺ニ構ヘ身ヲ捨テ体ヲ進メ夫レ突ヨト蕎フナリ相

手突出ス片刀炎々方ヲ打落シナガラ付込ミ左手ニテ小

手或ハ竹刀ノ柄ヲ押ヘ面部ヲ突トモ撃トモ為シテ勝ヲ

得ベシ又吾ヨリ乗込ンテ先々ノ勝ヲ得トスル片ハ對手

ト立合フハ片下段ニ構ヘ身ヲ捨テ乗込ムカ又ハ對手空ヲ

突キ蕗ヲ撃チテ構ヲ直ス頭ラヘ飛込ムカ何レ片敵ノ構ヘ

備ラザル所ヘ飛込ンテ勝ヲ得ベシ短劔ハ尺ノ短キヲ以テ

對手ノ劔ヲ應スルニ断理ナレ片手元ニ寄ラザレハ勝ヲ

The moment your opponent decides to attack, launch yourself towards *Temoto*, where his hands grip his sword and commit yourself to a *Sutemi*, Sacrifice Attack, in order to defeat your opponent.

If your opponent advances in *Jodan*, Upper Stance, you should take *Gedan*, Lower Stance. When he cuts down at your head, block with your bamboo sword and dart in towards his hands. With your left hand seize your opponent's wrist or the handle of his bamboo sword, thereby denying him freedom of movement before you stab him in the face.

If your opponent is advancing on you in *Gedan*, Lower Stance, then you advance with your sword hanging low in your right hand. You are advancing in a *Sutemi*, Sacrifice Attack, attempting to draw your opponent into attacking you with a sword thrust. When he stabs, use your sword to knock the tip of his sword down. Then dart in close and using your left hand to seize his wrist or the handle of his sword, before stabbing him in the face. By doing this you will achieve victory. If you are going to launch a pre-emptive attack, *Sen-Sen no Sho*, attacking and achieving victory before your opponent has a chance to formulate an attack, then begin in *Gedan*, Lower Stance. Leap towards your opponent with a *Sutemi*, Sacrifice Attack.

No matter how you confront your opponent, ensure that you force him to strike at air or otherwise react and cut at your feints. Having drawn your opponent out, you return to your basic stance and leap towards him to attack his head. Your overall strategy to victory is, no matter what position your opponent is in, leap towards where his stance looks weakest.

A short sword is a weapon without a lot of reach. When facing off against an opponent with a long sword, you must advance until you are beside where his hands grip the sword or you will not be able to achieve victory. Moving close enough to your opponent so that you are by his hands is a desperate move, that can be best described as "a life or death attack" and is therefore quite difficult to pull off.

得ルコトナシ手元ニ寄ルハ已ガ身ヲ捨ザレハ施シ難シ捨

身ノ術ニ付テ一話アリ一人ノ劍客遊歴シテ山中ニ至ル

幽谷ニ一本ノ圓木ヲ架シテ橋トナス其危險ニ望ンテ忽

チ疑懼ノ心ヲ生シ一歩モ進ム能ハズ石ニ腰シテ暫

時思慮ス時ニ一個ノ盲人此処ニ來リ杖ヲ以テ探リテ圓

橋ニ登リ杖ヲ搗ノ側面ニ跳テ、スラくト渡リ越キヌ是

ニ於テ劍客悟ル夫危懼ノ心アレハ進退遲滯シ捨身入身

ノ術施シ難シト后ケ一個ノ形ヲ造リ之レヲ圓橋ト擬セ

シト云フ

○巻藁土殴ノ辨

撃劍ヲ学フ者ハ日常巻藁土殴ヲ造リ置テ吾掌中ヲ試ミ

ルコ緊要ナリト曰常替古ノ上ニテハ分冶强ク勝ヲ人ニ讓

There is a little story about *Sutemi no Jutsu*, the technique of committing yourself fully to an attack where there are only two outcomes, victory or death.

There was once a swordsman who was on a pilgrimage, travelling all over Japan honing his sword skills. While travelling deep in the mountains, he came across a deep ravine. While there was a bridge traversing the gap, it only consisted of a single log laid across the span of the valley. Seeing the dangerous crossing the man was immediately struck by hesitation and doubt. He was rendered unable to proceed even a single step across the bridge.

Eventually, he sat down on a boulder and pondered his predicament. At some point a blind man with a walking stick came along the path. He walked up to the single log bridge and, tapping the sides of it with his stick, made his way smoothly across it. Seeing this the swordman had a great revelation, *If I have fear and hesitation in my heart and mind then I will be indecisive about advancing or retreating in any situation. This will make it difficult to execute Sutemi-Irimi, a sacrifice technique where you are thrusting your body forward to either win a duel or die in the effort.*

With this new understanding the swordsman developed a new technique called *Enkyo*, Round Wooden Bridge.

ラザルモ到底竹刀ヲ其人ノ仕癖ニテ掌中クルヒ平ニテ

打テ双方心付カスシテ勝ヲ定ルコアリ鎭劔ノ上ニ於テ

少シニテモ手裡狂ハ大ヒナル不覚ヲトルベシ因テ此

所ニ巻藁土段ノ図ヲ出シテ常ニ試ンコヲ勸ム

巻藁之圖
位
直径一尺

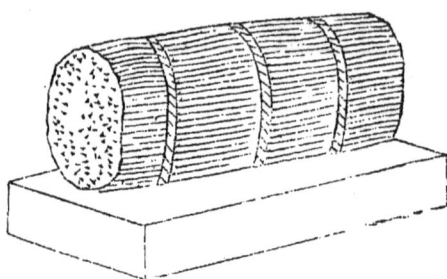

土段之圖
位
高二尺五寸

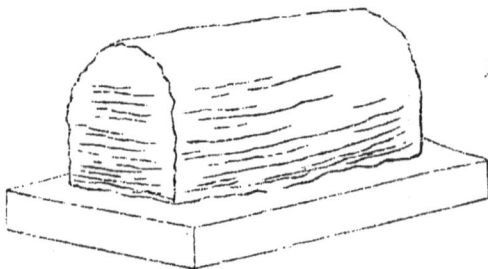

巻藁ヲ造ルニハ能ク乾キタル藁ヲ用ユヘシ濕気アレハ

ソクノ切レテ用ナキ物ナリ土段ハ海泥ニ砂ヲ好キ程

Why you should make a Straw Target and an Earthen Target

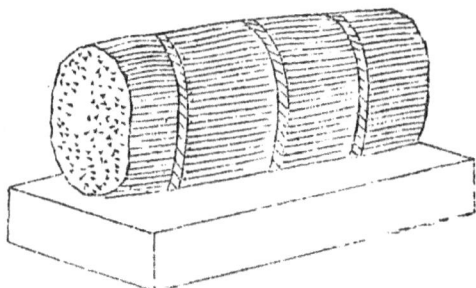

Makiwara, Straw Bale
Diameter 1 Shaku
1 foot/30 centimeters

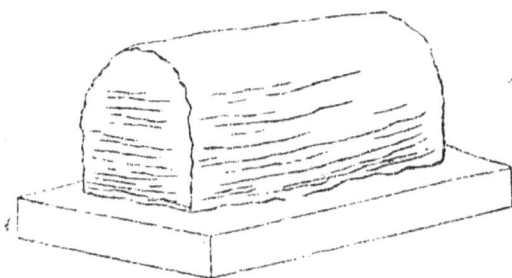

Dotan, Earthen
Mound
About 1 Shaku 5 Sun
in height 1.5 feet/45
centimeters

    Those that train *Gekken*, Japanese Style Fencing, or *Kenjutsu*, Sword Fighting, all construct *Makiwara Dotan*, Straw Targets and Earthen Targets. These are essential tools training grip strength. Keeping a firm grip on your sword is a fundamental part of training as it prevents your opponent from seizing the advantage and scoring a victory. In a duel, you are searching for where your opponent is vulnerable and striking in those areas. In other words, you are focused on achieving a clear victory, thus allowing your sword to twist in your hands may result in a strike with the side of the blade, instead of the blade itself.

    If you are in a duel with a *Shinken*, Real Sword, then failing to maintain control of the handle of your sword, even slightly, will lead to a perilous situation. Thus, an illustration of a *Makiwara Dotan,* straw target and earthen target, have been included important here and the author recommends training with them on a daily basis.

咬セ小石ヲ去リ圖ノ如クヌリ立テ少シ乾シ置キ和ラカ

ナル所ヲ切リ試ムベシ手裡ノ締リ悪シケレハ切レヌモ

ノナリ切リ方ハ外ニ子細ナシ両足ヲ踏揃ヘ上段ヨリ切

下ルナリ

　　　　　戸田流両分銅鎖之圖

戸田流ノ鎖ハ用法多ク傳ハラズト虫モ前三術ニ縁故ア

リ該世護身適当ノ具ナレハ使此ニ圖シテ同志ニ示ス其

用法大抵半棒ヲ使フ如シ

鎖分銅トモ鐵ナリ

文曲尺　　二尺

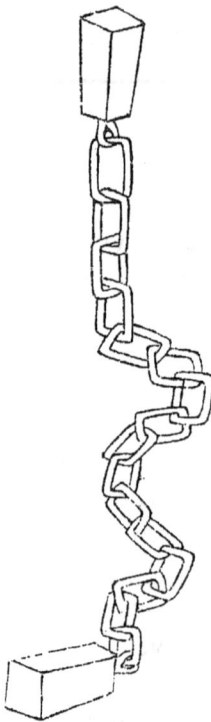

此畠ハ鎖分銅共鉄ニテ鍛工環ハ撫テ角ニシテ分銅ハ角

Kusari Fundo

Chain with a weight at each end, also known simply as Tetsu, Iron

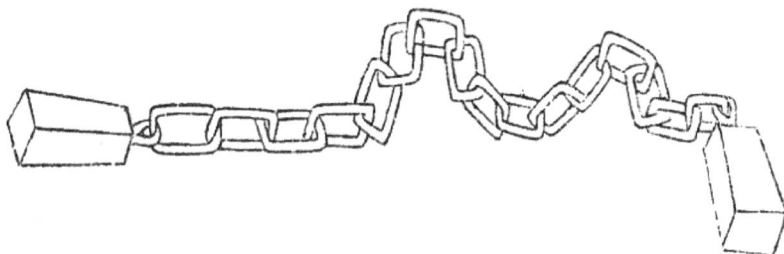

Length when extended, 2 Shaku 2 feet/60 centimeters

This illustration shows a *Kusari Fundo*, chain with a weight at each end, also known simply as *Tetsu*, Iron. The chain can be made out of rounded or squared rings. The weights at each end are rectangular prisms with clearly defined edges. Sew a case that matches the size of the weighted chain using high quality silk fabric, slip the chain inside and sew it shut. You can keep this within handy reach while you are at home and handle it like a toy. If you go out, the weighted chain can be tucked inside your belt still in its silk bag where it can be drawn out quickly to deal with a sudden situation. It can also be used to defend yourself against longer weapons.

ナリ縮緬ナドニテ細ク銷丈ケノ袋ヲ製シ之ニ入レ家ニ
在リテハ坐右ニ置キ常ニ翫トナシテ外ニ出ル片ハ袋ノ
佟腰ニハサミ須臾モ身ヲ放タザレバ変ニ臨ミテ長罰ニ
益アルベシ

遠当ノ法

遠当トハ所謂眼ッフシナリ水捕ノ具トモ云フ製法ハ左
ニ記ス所ノ水ヲ製シ瓶抔ニ貯ヘ置キ河豚ノ皮ニテ造レ
ル水砲制俚言キンタマツブシ夏季小児ノ翫フ河豚ニ入レテ袱ニ納メ置捕押者
等ノ時相手ノ面部ニ注ケハ忽チ眼ヲ閉ジテ開ク能ハズ

一 石灰ノアク水　　　右ノ水左ノ三味ヲ入

一 蕃辛細一松脂細一班猫タウガラシコ マツヤニコ 小
右加減ハ班猫ヲ入レサル前ニ試ミ適宜ニ制シ置ベシ

Toate no Ho
Striking From Afar

Striking From Afar refers to attacking your opponent's eyes in order to cloud his vison or distract him. It is also referred to as *Suitori no Gu*, Captured Water Tool. The next section will describe how to make this tool.

To make the water, mix one measure of red pepper with one measure of turpentine and a small, crushed tiger beetle. Before you add the tiger beetle, check how strong the mixture is and adjust accordingly. After mixing the liquid, store it in a jar or other such object.

Make a *Suiho*, Water Cannon, out of *Fugu*, pufferfish, skin. This is similar to the toy that children make in the summertime. These "water cannons" are colloquially called *Kintama Tsubushi*, Ball Crushers.

Fill the pufferfish skin with the above mixture and keep it in your sleeve. If confronted by an attacker, splash him in the face. This will prevent him from being able to open his eyes.

撃劍ハ右ノ手ニテ
鍔元ヲ握リ左ノ手
ニテ柄ノ頭ノ方ヲ
握リ右ノ足ヲ
進メ左ノ足ハ
趾タテ躰居コカズ
シテ自在ニ変化スル
ヲ考ヘシ余リ足ヲ
踏廣ケルハ悪シ、
上段ハ間隙多キ構
（ナル故ニスヘシ）
手ノ者ニ

一 對スル構ヘハ
星眼ハ對手
ノ目通リ切尖
ヲ着ケ下段ハ
胸部ヘ付ク
ルナリ

○應シ返シ
杯ト云テ劇キ
業ニ
テ間合
遠ケレハ
面部ハ打
ガタシ此時ハ
身ヲ引ナガラ
小手ヲ打ツ
ベシ

中段ニテ切リ結ラ
切リ返ヘロ

メ右ノ小手ヲ打
レタレバ左ノ片手ニテ
踏込ミテ面ヲ打
テ相打ニスベシ

274

When training *Gekken*, hold the sword with your right hand at *Tsuba-moto*, just below the hand guard. Your left hand holds the *Tsuka-gashira*, pommel. You advance with your right foot and your left follows. Your body should remain balanced so you do not stumble. You should be able to move in any direction freely in response to your opponent. Keeping your legs spread too wide apart is a bad habit.

Jodan, Upper Stance, is a posture that leaves you open to many different attacks, which is why you use it against weaker opponents. From that stance, thrust at *Seigan*, Star of the Eyes, or along your opponent's line of sight. If you are in *Gedan*, Lower Stance, then you will thrust to the center of your opponent's chest.

When attacking in *Chudan*, Middle Stance, you may end up crossed swords with your opponent. Here is a ferocious technique to counterattack from that position. If you have crossed swords with your opponent and find that it would be difficult to striking him on the head, instead drop back and strike him on the wrist. After striking him on the right wrist, release the sword handle with your right hand and, stepping forward strike him in the head one-handed.

此図ハ双方下段ニ
構ヘ折刀ノ切先ヲ
鶺鴒ノ尾ヲ動カス妙
ヲシテ透ヲ窺フ左ノ方ハ
突ヲ出サ下
ナス頭右ニ
落シ其儘
右ヨリ左ヘ
對手ノ竹刀ヲ拂ヒ
面ヘ打ヨリ
十分ノ勝ヲ
得ル

是ハ右ノ方
上段ニ構ヘ
左ハ星眼
二三テ立合
ヒ左ハ体ヲ
カシ左足ヲ
踏込ナカラ左ノ
二手ニテ面ヲ突
キ勝ヲ取ル
此時相手モ早
ク身ヲカワシ左
足ヲ踏込
二面ヲ打
ミ片手打
又右手ヲ
引ナカラ小手或ハ
竹ヲ打チ落シ面ヘ掛ルモ
亦妙ナリ

This illustration shows the duel described below. Both you and your opponent begin in *Gedan*, Lower Stance. While looking for an opening, you build striking power in your *Shinai*, Bamboo Training Sword, by flicking it up and down rapidly like the long tail of a Japanese wagtail bird searching for insects on the beach. When your opponent on the left appears ready to launch a *Tsuki*, Straight Thrust, shift your body from right to left, knocking your opponent's sword aside. Then launch yourself forward and strike him in the head, achieving the win.

In this duel, your opponent on the right is in *Jodan*, Upper Stance, while you, on the left, are facing him in *Seigan*, Starry Eyed Stance. Step forward with your left foot, rotating your body clockwise, and by stabbing your sword forward with just your left hand, you can strike him in the face and achieve victory.

If your opponent is quick, he can dodge by stepping forward with his left foot at the same time, dodging your attack. He can then follow up by striking your head or face with a one-armed attack. He can also pull his right hand back and strike your wrist or, if he is an advanced practitioner, knock your bamboo sword aside and attack your face.

是ハ相下
段ニ切結ヒ
互ニ合一方ハ間ヲ計リ
テ踞嫩右胴ヲ打チ
勝占メタリ

丸胴ハ相手ノ右ノ方
ヲ打ツベシ然レ匕胴
打片ハ巳ノ面明キテ甚
危シ且手ノ返リ方悪
シケレバカノ平甚難キ
伎ナリ真剣ニ殊ニ用
ヒニクシ注意アルベシ

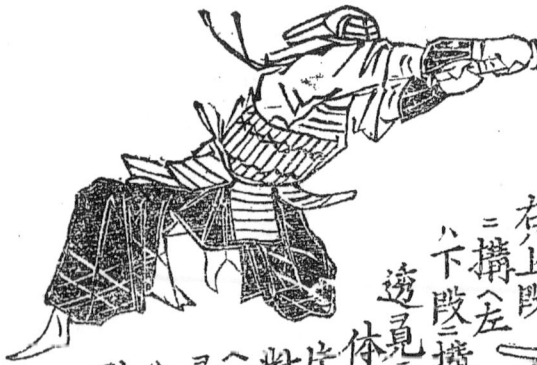

右上段
ニ構ヘ左
ハ下段ニ構ヘ
遊見テ右ハ
体ヲ開キ
片手打ニ
對手面
ヘ打込ハ
一方
ハ
引方ヲ
身ヲ
引テ
小手ヲ
打テ勝ヲ得タリ流義ニヨリテヘヲカケ小手
又引小手ト云ヘリ真劒ニテ

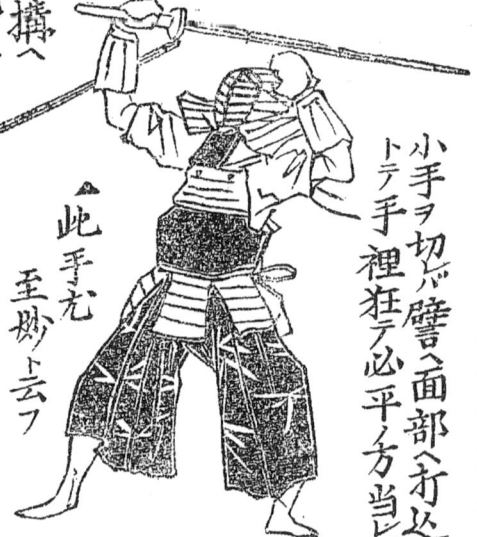

右上段
ニ構ヘ左

此手尤
至妙ト云フ

小手ヲ切バ譬ヘ面部ヘ打込タリ
トテ手裡狂テ必平ノ方当レバ

278

In this technique both you and your opponent are in *Gedan*, Lower Stance. You have both struck simultaneously and ended up with your swords crossed. The moment you break apart, drop to your left knee while keeping your right upright and strike your opponent in the right side. This will clinch victory for you.

When attacking a person's abdomen, only strike your opponent's right side. However, you leave your head and face wide open when attacking an opponent's side, which is dangerous. Further, if you do not properly rotate the blade of your sword when cutting horizontally, you will end up striking with the side of your sword and the cut will not be effective. Be aware that this technique is very difficult to use with a real sword.

In this technique your opponent is on the right in *Jodan*, Upper Stance, while you are on the left in *Gedan*, Lower Stance. Detecting an opening, your opponent rotates counterclockwise, attempting to strike you in the head by releasing his left hand and cutting with the sword in just his right hand. You respond by dropping back and striking him in the wrist, thereby winning the duel.

Depending on the school, this type of attack is called *Kake Kote*, Wrist Cut When Attacked, or *Hiki Kote*, Pulling Back Wrist Attack. If you cut your opponent's wrist with a real sword duel, you would sever the arm and cut into your opponent's head. However, your sword may twist in your hands, resulting in a horizontal strike. There are many difficult and mysterious aspects to this technique.

短刀用法

長刀
上段
ニ構ヘ
面ヘ
打コマニ
スルトキハ
短刀ヲ下段ニ
構ヘ打込ム
竹刀ヲ受
ケ流シナ
カラ手元
ヘ飛入左ノ
手ニテ對手ノ
中柄ヲ取リ面ヲ突

護身ノ構

長刀星
眼又ハ
下段ニ
構ヘタルトキハ
短刀ヲ提ゲ体ヲ
スメテ突ケヨガシト
隙ヲ見セ欲ハ突
カントスルトキ短
刀ニテ巻カヘル
ヤウ物打ヨリ
火先ノ方
キリキリ
ヘ打拂
飛込ニ
對手ノ左
右小手ノメ

捨身ノ構

×
間ヘ左ノ手ヲ
差込ミ自由ヲ
拗ケ面ヲ打

How to Fight With a Short Sword

Goshin no Kamae
Self-Defense Stance

Your opponent is armed with a Long Sword and is standing in *Jodan*, Upper Stance. The moment he moves to attack your head, shift to *Gedan*, Lower Stance. Block and pass your opponent's attack while darting in towards where your opponent is holding his sword. Seize the center of his sword handle with your left hand and stab him in the face with your right.

Sutemi no Kamae
Sacrifice Stance

Your opponent is armed with a long sword and is standing either in *Seigan*, Starry Eyed Stance, or in *Gedan*, Lower Stance. You advance towards your opponent with your short sword held low, inviting him to attack. Thinking that you are wide open, your opponent launches a *Tsuki*, Straight Thrust. You respond by striking the tip of his long sword as if trying to wrap up his sword, sweeping it away while moving in. Having closed the distance, slip your right left hand between your opponent's hands denying him freedom of movement while striking him in the head with the sword in your right hand.

282

長刀面ヘ打込
片真直両手ヲ
延ハシ差上ケ
左ノ手ヲ申ト
短刀ノ右ノ平
ト間ニ受留メ
長刀ヲ引カント
セバドコマデモ
付入リ虚ヲ窺
左ノ方ヘ受流シ
ナカラ付入左ノ
手ニテ中柄ヲ
執リ面ヲ突
ナリ

鍔取

長刀下段
ニ搆ヘ伎
ヲ施コサズ
隙ヲ扮
ヲモ居片

試ミニ短刀ヲ
長刀ノ左ノ平ヘ
着ケ敵ノ動
静ニ因テスリコミ
勝ヨリル

口橋

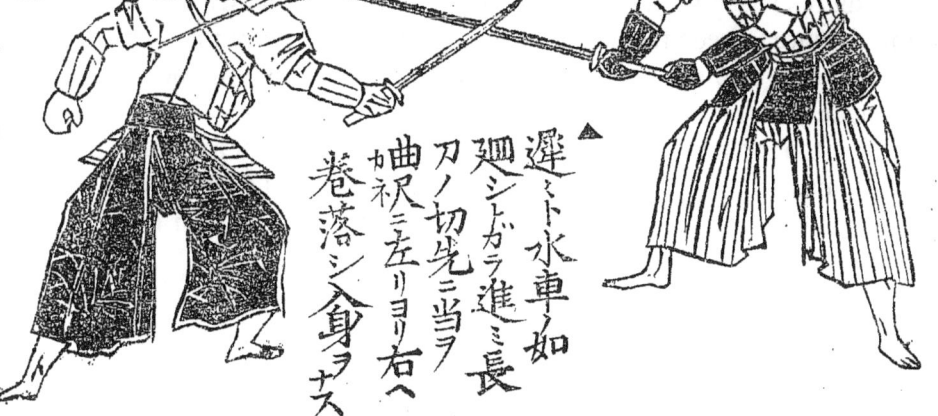

遅ミト水車ノ如
廻シ上ガラ進ミ長
刀ノ切先ニ当リ
曲尺ニ左リヨリ右ヘ
巻落シ身ヲ
ス

水車
水車ハ
遠ク
隔テ
短刀ヲ
右ヨリ
左ヘ

Tsuba Tori
Seizing the Hand Guard

In this technique your opponent is armed with a long sword and he is striking at your head. Respond by extending both arms upward and blocking his strike with your short sword held horizontally out to the right. Be sure to keep the back of your left hand under your right hand supporting it. The moment your opponent pulls his sword back, you will make your move.

Judge the moment and feint, before shifting to the left and moving in while passing his sword to your right. Seize the center of his sword handle with your left while stabbing him in the face with your right.

Round Bridge

Your opponent is in *Gedan*, Lower Stance, but is not attempting any technique. Instead, he is looking for an opening. Respond by placing your short sword on the left side of his long sword, and wait for him to respond. If he does not, slide in and strike, achieving victory.

Water Wheel

Water Wheel is when you are armed with a short sword and start a good distance away from your opponent. Moving slowly counterclockwise you advance on your opponent like a rotating water wheel. You then extend your sword towards the tip of your opponent's sword from and with a circling motion wrap his sword up in a counterclockwise motion and force it down, before advancing to striking range.

棒ハ流儀ニ依テ多敷有

又十八手卅手

外様々々ナリ

故ニ爰ニハ

聲セス

此図ハ

棒ノ振法ヲ

記ストモ甲乙ニ分

最初ハ双方図ノ

如ク身搆ヲナシ

掛声應シ双方

共ニ第二圖ノ

如ノ立ナリ

双方右

棒ノ手ニテ

棒ノ先ヲ

握リ互ニ

ツ、ク

○

（二）半身ニ搆ヘ（左ノ

手ニテ棒ノ中央

ヲ握リ図ノ如ク

突立ナリ

（ヤイヤ）ト云テ

下テ打合テ

三度ナス（左右トモ）

一ツ都テ棒ハ中央ノ

所ヲ握ルナリ

棒ハ振様ニシタカヒ

右ノ手先ニシタカヒ

時ハ我躰モ

右足ヲ先ニシ

総テ体ニ付テ

揮廻スナリ

（棒ヲ振替持替

ル時ハ此插画ノ如ク

都テ中央ヲ持

替ルナリ

両手ヲ

指ヲ延

シテ合

掌ノ如

ク手ノ

ハコビヨキ

ヤウニ

スナリ

Bojutsu
Wooden Staff Fighting

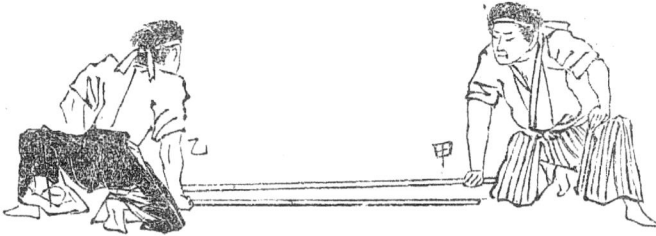

There are many different schools of *Bojutsu*, wooden staff fighting. The techniques can not all be covered here as some have eighteen techniques and others have thirty techniques.

The illustration above shows the proper etiquette before training. The two combatants are labeled *Ko*, former and *Otsu*, latter.

Initially, both you and your opponent squat in the positions shown in the first illustration. [35]

Next, both stand and hold the staff in the middle with your left hand and the end with your right. Stand with your side facing your opponent. This is shown in illustration two. Simultaneously you both do a lower strike while shouting *Eiya!* Do this strike three times (the strike should be done on both the left and right sides.)

When doing Bojutsu, you always hold the staff in the middle. If you are having trouble rotating with the staff, remember to also step forward with your right foot when extending your right hand. This means your body will rotate counterclockwise along with the staff.

To switch the staff from one hand to the other, ensure your hands are together as if in prayer, with your fingers extended. The above illustration also shows how to flip the staff around and use the back end as the front.

---

[35] The color and pattern of the combatants' clothing changes every scene and does not consistently represent one or the other.

甲モ
図ノ如
クニシ左
足ヲ後ヘ
引キ半
身ニカマヘ
右ノ手ニテ
突來ル棒ヲ▲

乙ヨリ（ヤー）ト声ヲ掛テ
体ヲ半身ニ構ヘ第四図ノ
如クニナシテ甲ノ胴ヲ突ク

一寸ト
立

棒ヲ両手
ニテ握リ
甲乙共ニ

双方左右左下ニテ
打合テ第三図ノ
如ク七分三分ニ

第
三
圖

乙

甲

▲体ヲ
クルリ
ト後ヘ
廻リ
乙ニ空
ヲ突
セテ
左ノ手
ヲ持チ
添ヘテ廻ル途端ニ
乙ノ棒先ヲ打排
フナリ

第
四
圖

甲

乙

Illustration 3

After both combatants practice striking low three times: first left, then right and left again, they shift to the stance shown in the illustration above. Hold your staff with both hands so 70% of the staff is below your hands and 30% is above your hands. Having taken this stance, you both pause.

Illustration 4

Your opponent shouts *Yaa!* and steps back with his right foot rotating clockwise and stabs the end of his staff at your abdomen. This is shown in the above illustration.

You respond by stepping back with your left foot and rotating your body counterclockwise. You are now positioned with your right side facing your opponent and your staff in front of you held in your right hand. By rotating back away from your opponent, the end of his staff strikes nothing but air.

Next, join your left hand to your right on the staff and, while rotating it around, strike the end of your opponent's staff with a hard, sweeping motion.

第五圖

前ノ手ニ次キテ棒ヲ拂ヒ

直ニ又甲ヨリ（エイヤト

右ノ手ヲ差延

乙ノ
×

乙

甲

●体ヲ替乙ノ足ヲ拂フ乙ハ

足ヲカヽメテ外ツス

リ

第

六

圖

×真向ヘ

打落ス

乙ハ（ヤレト答

ヘテ流シ棒

両手ニテ甲ノ打下ス所受止甲又ハ（エイト

乙

甲

Illustration 5

Your opponent sweeps away your previous attack. You immediately follow this by shouting *Eiya!* and extending the staff with your right hand, aiming to strike down on the top of your opponent's head. Your opponent responds with a shout of *Yaa!* and swings his staff around before forcibly knocking your staff downward, thereby blocking your strike.

Illustration 6

You shout *Ei!* and rotate your body around aiming a sweeping strike at your opponent's legs. Your opponent dodges by lifting his legs up as shown in the above illustration.[36]

---

[36] The illustrations do not seem to fully match the descriptions, the author and artist may have been trying to convey different types of movement.

第七圖

乙ハ我ガ体ノ入
替ルヤイナ
（エイと云テ
右ノ手ヲ先
（出シテ
打込ム×

甲×
八足
ヲ引
クト
時ニ流
ヒニ棒ヲ
体ニ付テ
ひノ打込ム
⦿

棒ヲ体ヲ半身ニ
入替テ受留
ルナリ
一図ヨリ
八図ニ至ル
マテ一組
トナス
ナリ
⦿

第八
圖

Illustration 7

Illustration 8

With a shout of *Eiya!* your opponent rotates counterclockwise and extends his right hand striking at your head. You respond by pulling your foot back and, at the same time sweeping your staff across his feet. Then pull your staff against your body, drop back while rotating counterclockwise and block your opponent's follow-up strike as shown in the illustration. The illustrations up to this point represent one sequence.

棒ト木太刀ノ法

甲

棒ノ法ハ
都両三尺ノ隔ヲ
要務ト心得フ
ベシ

一木太刀棒ノ立合ハ三四尺ヲ隔ツヲ
定法トス偕双方自眼合乙ハ太刀ヲ真向ニ

乙

第二圖

撃シテ
打セル
甲ハ図
ノ如ク
名棒
突立
我カ躰ヲ
引上
後ヘ反ル
又乙ハ
二度
ウツ

方込ム
甲父壱足下リ
体ヲ又替第三
図ノ如ク

乙

Dueling Sword Versus Staff

Remember when dueling with a staff that you should always begin 3 Shaku, 3 feet/ 90 centimeters, away from your opponent. When dueling against an opponent armed with a *Kidachi*, wooden long sword, the rule is to begin 3~4 Shaku, 3~4 feet/0.9~1.2 meters, apart. Both you and your opponent should be glaring at each other fiercely.

Your opponent attacks, aiming to cut straight down on your shoulder. You respond as shown in the illustration by pulling the end of the staff you had planted in the ground backwards and up while leaning back. Your opponent continues to advance and strikes again.

第三圖

乙ノ打込手首
振返シ
タル棒ノ先
ニテ左ノ方ヨリ
打落ス

第四圖

勢ヲナス乙ハ
其棒ヲ受止メ
又甲ハ棒ヲ振替
体モ入替
一足進ミ
右ノ方ヨリ
右ノ足ヲ掃ツ
乙ノ勢ヲナス
乙ハ又我体ヲ一足下リ
テ受止メ甲ノ体入替ラヌ内ニ

×
ヒヨリ(ヱモ)ト真向ヘヤ又
打込勢ヲナス

第九ヨリ
十三至
棒太刀
ノ音合
一組トス

You respond as shown above, by dropping back with your left foot and rotating your body counterclockwise. As you do this swing your staff from left to right, aiming to strike your opponent's wrists as he cuts, forcing his arms down.

Your opponent blocks your staff with his sword.

Next, flip your staff around while shifting your weight forward and stepping forward one step. Swing your staff from right to left, aiming to strike your opponent in the right leg. Your opponent avoids this by dropping back one step and blocking with his sword. Before you can drop back, your opponent shouts *Eiya!* and steps in cutting straight down at your head.

Illustrations 9~12 represent one staff versus sword technique.

半棒ノ圖

俗ニ半棒ハ三尺棒ヲ×

×云フ甲乙共三尺
余ヲ隔テ對立使用
スルモノニシテ六尺棒ト
同様成レトモ

▲セマキ所ニテハ最
ニテハ最
便ナリ至テ妙ノ
手敷多アリ○

○胸膈ヲ
突カントス甲ハ
圖ノ如ク身ヲ構
ヘ突出ス棒ヲ

○現図ヲ好々見合セ順序ニ振ルベシ
第一図ヨリ六ニ
至ルヲ二手トナス
乙ヨリ甲ノ○

○両手ニテ〈ヱ卜ヲ下落
ス其時直ニ我体ヲ
半身ニナシ右後ニ足ヒキ
上段ヨリ乙ノ直向ヘ
打込ム※

※乙モ棒
先ヲ左ヨリ横ヘ
流シ打ニ拂フベシ

Hanbo, Half Staff

Hanbo, Half Staff, is the colloquial name for the *San Jaku Bo*, 3 foot/90 centimeter staff. Both you and your opponent are facing off about 3 feet apart. The weapons you are both holding are basically half of a standard six-foot wooden staff, however they are ideal for dueling in confined spaces, thus there are many interesting techniques.

Look carefully at this illustration. Illustrations one through six will show two techniques.

Your opponent will attack first by stabbing at the center of your chest. You are positioned as shown in the illustration. When your opponent strikes you shout *Ei!* and drop both hands down. Then immediately drop back with your right foot while raising your half-staff into *Jodan*, Upper Stance, and cut down straight at the top of your opponent's head. Your opponent responds by swinging the end of his staff horizontally from the left to sweep across your middle.

又壱手双方共打込手ニテ
ナリ甲乙棒先ヲ握リ
（ヱ、ト、曲）

棒先ニテ打合ス

云サマ乙ノ真向ヨリ
打落ス乙ハ（ヤ、ト）両
先ヲ握リテ受

留メ直ニ甲ノ胴ヘ打込ム甲ハ我カ体ヲ棒ニ付テ
半身ニ構ヘ受留ナリ以上外数手有後篇ニ詳記スシ其他長刀鑓六尺棒寺
釼彔術中段極意秘傳併セテ細出ス或後篇ヲ待テ詳解シタマヘ

The next technique begins the same way as the first one. Both combatants strike at the same time and the ends of both strike at the same time.

Both you and your opponent are gripping the ends of your weapons in the same fashion. You shout *Ei!* and strike straight at the top of your opponent's head. Your opponent responds with a shout of *Yaa!* and blocks your attack by holding both ends of his staff.

He then immediately strikes horizontally to your torso. You defend against this by rotating your body clockwise and holding your staff vertically against your body.

This ends the section introducing several staff fighting techniques. More details will be available in the next volume, which will contain halberd, chain and sickle, wooden staff fighting and more. [37] In addition, it will include a great many mid-level and advanced sword and Jujutsu techniques in addition to other material, so readers are advised to look out for this volume.

---

[37] The second volume was released in 1890

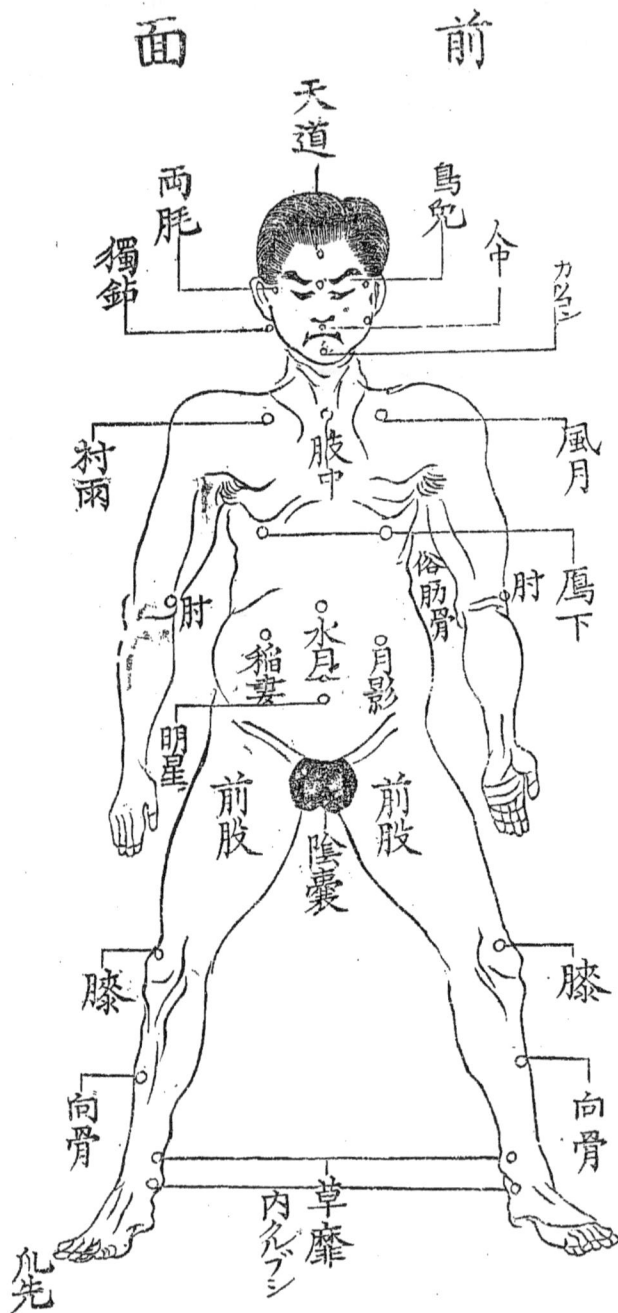

當身之真圖

面　前

爰ニ図スルモノハ只当身灸所ヲ記シテ其名称所ヲ記憶スルノ為ニセルノミ
初心ノ輩活法ヲ不知シテ妄ニ施スカラス其詳ナルハ後篇ヲ待テテ了解シ至ヘ

天道
鳥兎
両胝
令
獨鈷
カクシ
村雨
股中
風月
肘
鳫下
俗肋骨
肘
水月
月影
稲妻
明星
前股
前股
陰嚢
膝
膝
向骨
向骨
草靡
内クルブシ
爪先

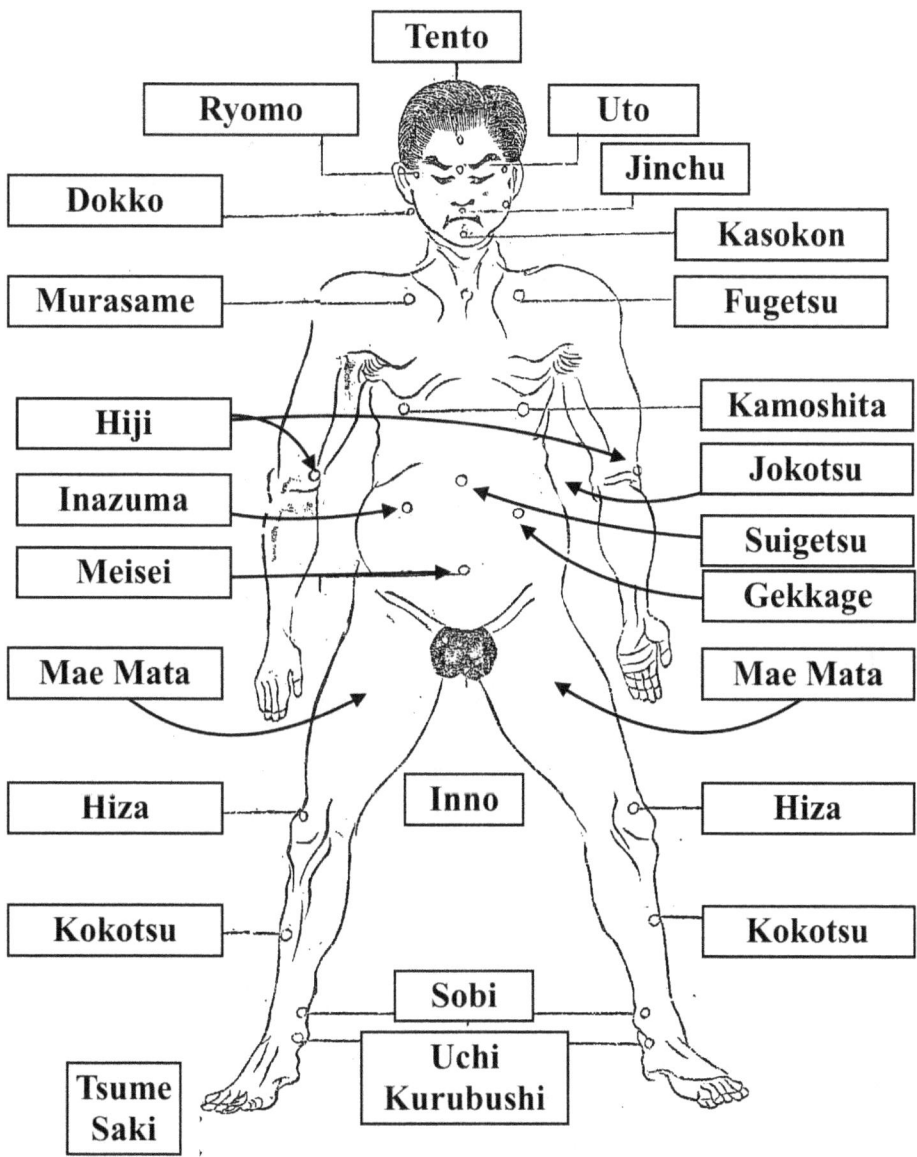

Tento

Ryomo

Uto

Jinchu

Dokko

Kasokon

Murasame

Fugetsu

Hiji

Kamoshita

Inazuma

Jokotsu

Suigetsu

Meisei

Gekkage

Mae Mata

Mae Mata

Hiza

Inno

Hiza

Kokotsu

Kokotsu

Sobi

Uchi
Kurubushi

Tsume
Saki

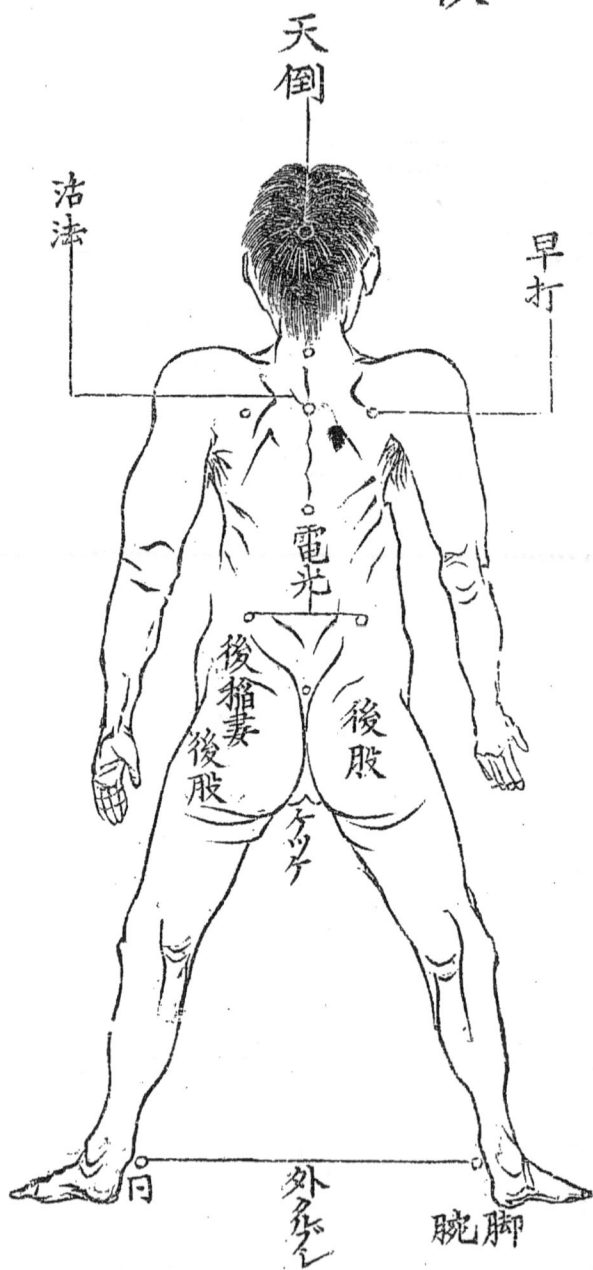

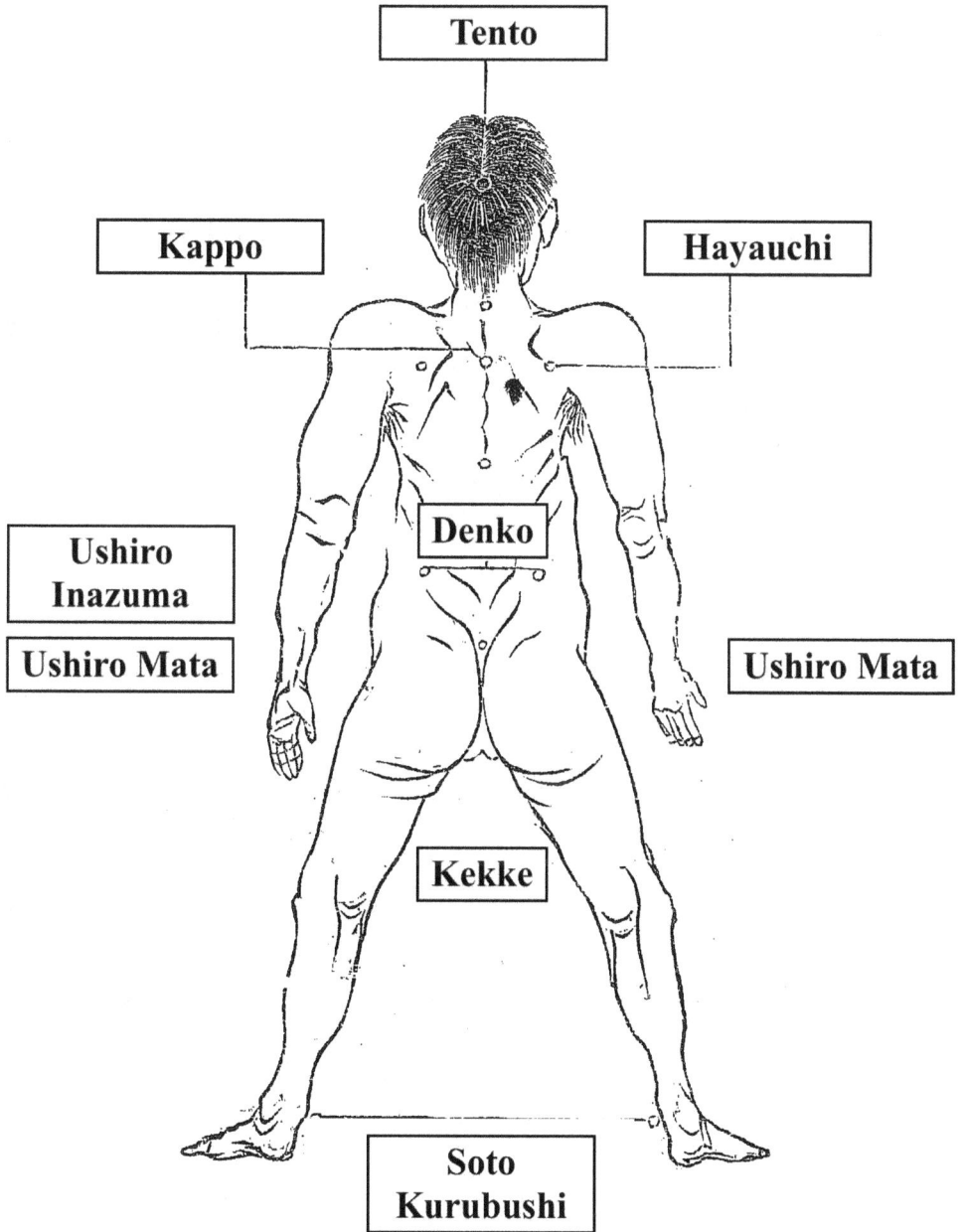

Tento

Kappo

Hayauchi

Denko

Ushiro Inazuma

Ushiro Mata

Ushiro Mata

Kekke

Soto Kurubushi

早繩心掛ノ事

早繩ハ三尋半本繩ハ七尋半十一尋ヲ用ユ早繩ニハ後圖
ノ如キ鈎ヲ付クルヲ善トス捕繩ノ製ハ麻ニテ三ツグリ
ナリ光澤カナルヲ用ユベシ此術ヲ心掛ルノ主意ハ亂暴
人發狂人又ハ盜賊搦捕押ヘタルトキ先ツ其手足ヲ縄メ
自由ヲ得セシメズ然ル後其筋ニ訴フ可シ

鈎繩ノ圖

犬ケ寸七分位

早手錠ノ畧

大小隨意
分銅堅キ
木ニテ造ルベシ

Haya Nawa Kokoro Kake no Koto
Lesson in Quick Tie

Haya Nawa, or Quick Tie, uses a rope 3.5 Fathoms, 20 feet 8 inches/ 6.3 meters in length.[38] This is also known as *Kukuri-han*, A Wrap and a Half. The rope used for Hon Nawa, Main Tie is 7.5 ~ 11 Fathoms, 44~65 feet/14 ~ 20 meters. The 7.5 Fathom rope can also be referred to as *Nana Tatari Han*, Seven and a half divine punishments.[39]

Your *Hojo* Arresting Rope, also known as *Tori Nawa*, Seizing and Tying Rope, should be constructed of three strands of hemp rope twined together. It is best to use soft and flexible rope.

You should learn rope tying techniques so that you are prepared to restrain and tie a violent person, a person who has lost their mind as well as burglars.

The first thing you want to do is remove the criminal's freedom of movement by tying their hands and feet. Only then should you make your complaint to the authorities.

---

[38] A *Hiro* 尋, fathom, is an old unit of measurement equivalent to 6 feet or 180 centimeters. The same word can also be read as *Jin* 尋. A length of rope 3.5 Hiro is 20.6 feet.

[39] *Tatari* 祟り Refers to being cursed or receiving divine punishment from an angry spirit.

犬ケ寸七分位 鈎縄ノ圖

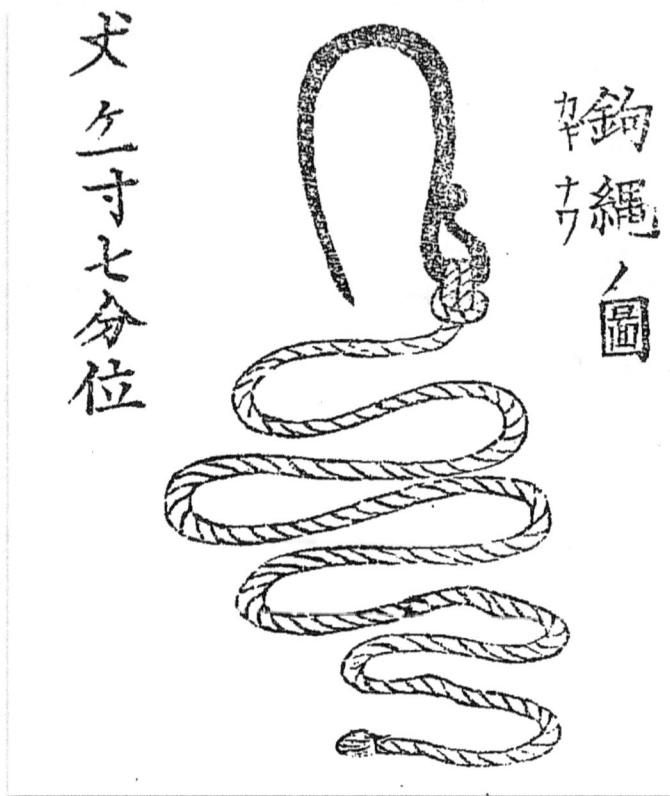

Kagi Nawa no Zu
Illustration of a Kagi Nawa Hook and Rope

You can also attach a hook to your Quick Tie rope as shown in the illustration. The hook should be made out of sword steel and crafted in the shape shown in the illustration. You should forge the hook carefully so that it does not bend.

The width of the opening at the bottom of the hook should be 1 Sun 7 Bun, 2 inches/5.1 centimeters

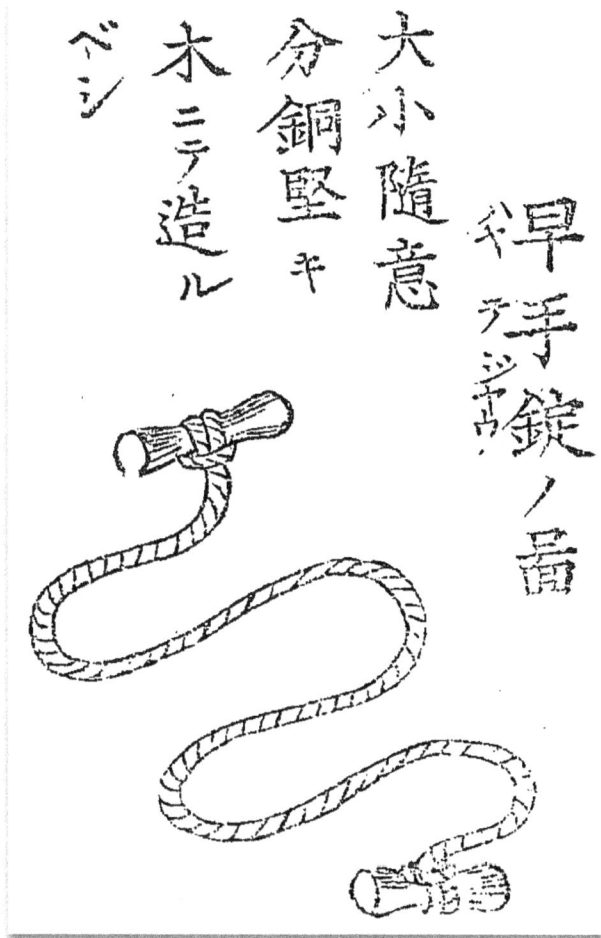

早手錠ノ図

大小随意分銅堅キ木ニテ造ルベシ

Haya Te-Jo no Zu
Illustration of Quick Handcuffs

The weight at the end should be made out of a hardwood. The ends of the wooden pegs at the ends of the rope can be made larger or smaller, according to your liking.

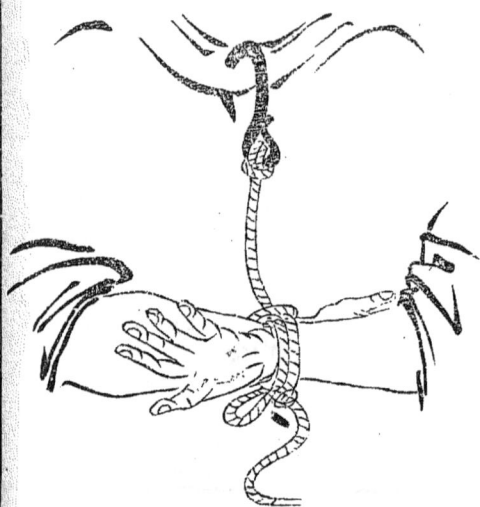

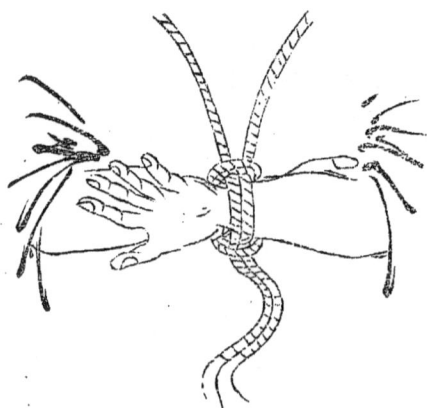

鉤ハ叙ニテ圖ノ如ク錢ビザル様ニ能ク鐵ヘ造ラスベシ

用法ハ對手ヲ捕押ヘタル片此鉤ヲ襟ニ通シテ小手ヲ溜

ムベシ其他對手ヲ捕押ユル片其摸様ニ因リ衣類又ハ疊

建具何ニテモ手近ニテ鉤カナル物ニ鉤ヲ打込ミ其繩ヲ

徹ノ足手道抔ヘ繋ケテ引倒ス可シ大ニ切用アリ

The hook can be used in various ways. When you are seizing and restraining a person you attach this hook to his collar. You then use the rope to secure his hands. Depending on the situation you may have to attach the hook to other parts of their Kimono or even secure the hook to a Tatami mat or other part of a building. After attaching the hook to something secure, you can then use the rope to secure the hands, feet and neck of your enemy. Pulling them down on the ground will make this more effective.

小手ヲ締ルニハ畾ノ如ク左右ノ手ヲ後ロニ抱キ合セ手

甲ト節トノ間凹ナル所ヲ二タ巻キ廻シ夫ヨリ右左ノ手

ヲ割リテヌ二タ巻廻シ夫ヨリ下ノ方ニテ垣根結ビニ留

ルナリ

早懸ノ畾

垣根ムスビ

同

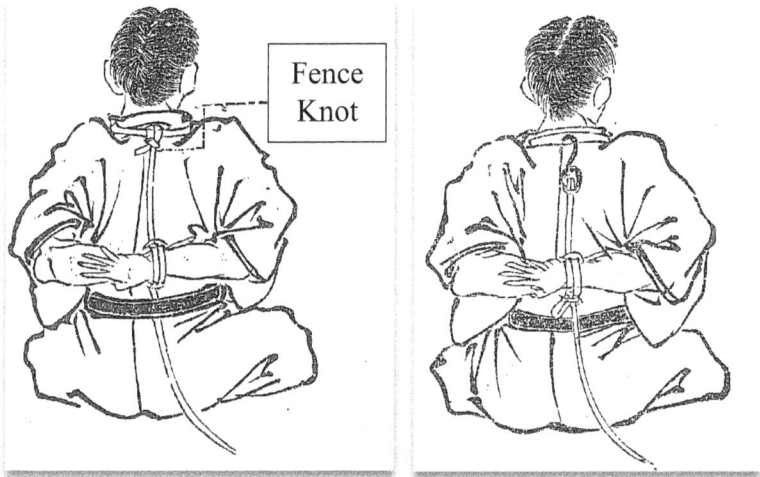

Haya Kake no Zu
Illustration of Fast Tie

Secure the hands by tying as shown in the illustration. Bring both his hands behind his back, making sure each hand is holding the wrist of the other.

Wrap the rope twice around the wrists at the point where there is a slight depression at the base of the hands. Then wrap the rope twice between the two hands and secure at the bottom with a *Kakine Musubi*, Fence Knot.[40]

---

[40] Kakine Musubi, Fence Knot. Other names for this knot include Otoko Mususbi (Men's Knot,) Moro Musubi (Double Knot,) Ibo Musubi (Wart Knot,) An Mususbi (Hermiatage Knot,) Shiori (Garden Gate Knot) and Hae-gashira (Fly's Head Knot.)

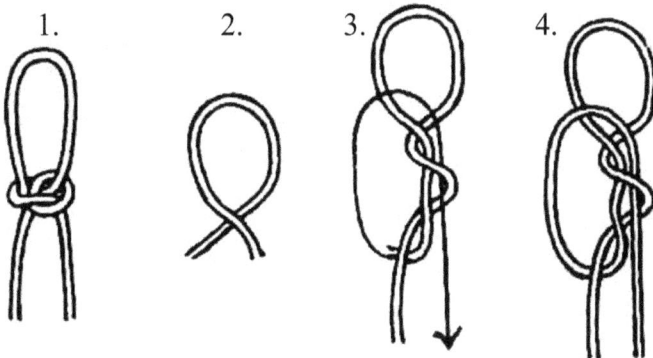

1.   2.   3.   4.

早縄ノ図

本縄

菱縄ト

袷入

カマガクシ

同

Hayanawa no Zu
Illustration of Hayanawa

Hon Nawa, **Main Tie,
also known as** *Hishi
Nawa,* **Water Chestnut
(Diamond Shaped) Tie**

Kama Gakushi
Hidden Sickle Tie

Another version of
Hon Nawa

本縄ハ數種アレ𪜈唯其一種ヲ示ス縄ハ七尋半ノ麻ヲ用ニ繒ノ

縛リ襷ハ七尋半ヲ新キ垣根結ヒニナシ夫ヨリ縄ヲ左ノ

右ニ吩ケ先ッ左ノユノ腕ヲカマガクシニ掛ケ又右ノ二
ノ腕ヲ同樣ニ掛ケ右左ノ縄ヲ一ッニ集メ帯ノ上ノ所ニ
テ引キク、リノ輪ヲ出シ置キ扨兩手ヲ前﨑ノ如ク抱キ
合セ縄ノ末ヲ引括リノ輪ニ通シテ輪ヲ引括リ縄ノ末ヲ

左右ニ分ケテ兩手ノ間タヘ割込ミ一ト廻シ廻シテ垣根
結ヒニ留其縄ノ末ヲ帯ヘ通シ置ク可シ

浮手鏁ハ前圖ノ如ク手杵樣ノ物ヲ鏨キ木ニテ造ル大サ
曲澤二寸位ナリ之ニ細キ縄曲尺一尺七八寸ヲ付シ置ナ
リ調法ハ早縄銅樣ニ小手ヲ巻キ兩手ノ間ヘ割込ミ締メ
タル末ヲ幾囘モ捻リ吩銅ヲ小手ノ間タヘ挾ミ置クナリ

While there are a great many variations of the Hon Nawa, Main Tie, only one will be introduced here. The rope is 7.5 Fathoms, 44 feet long. To do this tie, you will first need a loop for the neck. Make this by folding the 7.5 fathom rope in half and tying a fence knot.

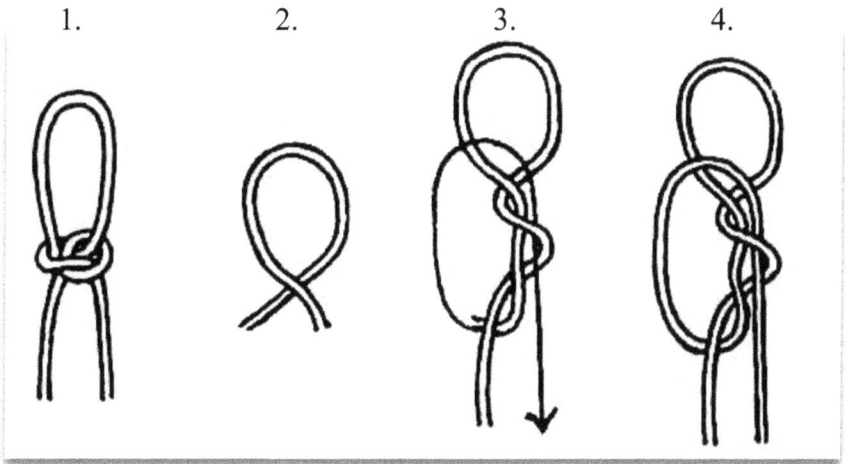

After tying that knot, take the left strand and tie a *Kama Gakushi*, Hidden Sickle knot around his left arm. Then tie the same knot around his right arm with the right strand.

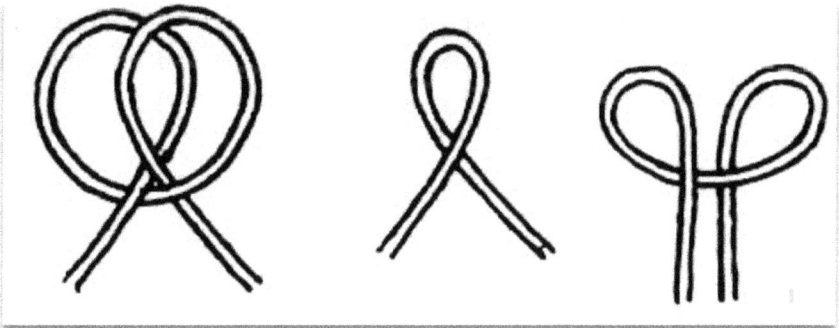

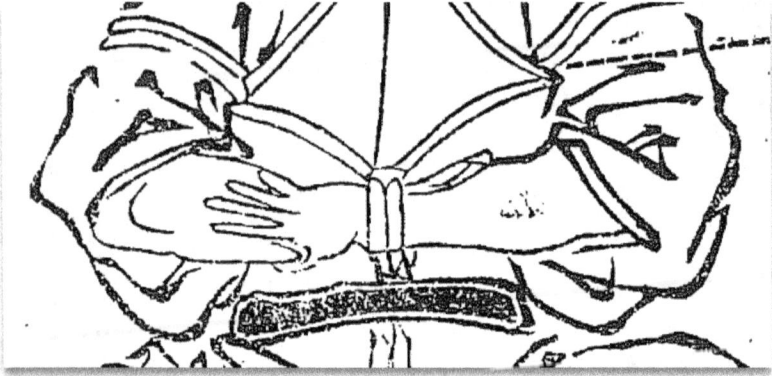

Bring both ends of the rope towards the prisoner's waist and wrap them around the back of his belt. Then make a slipknot and insert both hands as shown in the previous illustration. Each of the prisoner's hands should be holding the opposite wrist. Pull both ends of the slipknot, tightening it around his wrists and then take both ends of the rope. Wrap the ends of the rope around in between the hands first one time, then a second time. Finally, tie a fence knot at the bottom and thread the remaining rope through his belt.

The illustration above shows *Hase-te Jo*, quick rope handcuffs. They should be made out of a hardwood and should be about 2 Sun, 2.4 inches/6 cm long. Use a ruler to measure them. Also use the ruler to measure 1 Shaku and 7 or 8 Sun, 20~21 inches/51~54 cm, of thin rope. After attaching the rope, the handcuffs are finished.

Quick Rope Handcuffs are used in a way similar to Quick Tie. Wrap them around both wrists and then in between the hands. Secure them by twisting the two ends around several times and wedging the weighted ends between his hands.

This will keep your opponent secure for a short amount of time. When it comes time to move your prisoner however, you should apply Main Tie.

之ハ全ク瞥時ノ廻ノナリ時間ヲ移スコトアラハ本縄ニ掛

代ユヘシ

三寸縄五寸縄ノ大変

三寸縄五寸縄ノ朝法ハ秘後ナリ之レ別法ニアラス雖

三寸五寸ノ物ヲ以テ人間五尺ノ体ヲ縛シ彼ヲシテ自由

ヲ得セシメサルノ法ナリ對手ヲ捕ヘ其両手ヲ本人ノ襟ニ

引揚ケ両手ノ肱ヲ組ミ合セ切元結

ニテ圖ノ如ク廻ヲ元結ノ末ヲ襟

ニ捥ヶ瘠ヶ通シテ結ビ置ケハ指

ヲ拔ク能ハス又編ミテ引切ルコヲ得

サル可充元結ニ限ラス三縦糸ナト誼シ只尺ニ満タサル

物ニテ人ヲ縛スルノ妙ヲ示スノミ

San Sun Nawa Go Sun Nawa no Daiji
The Secret Teaching of the 3 & 5 Sun Rope

There is a way to use the 3 Sun, 3.5 inch/9 cm rope, and the 5 Sun, 6 inch/15 cm rope, which has been kept secret. There is no other method like it. With just a 3~6 inch-long length of cord this method allows you to restrain a 6 foot tall man, preventing him from moving.

After capturing your opponent, twist both arms behind his back and pull his hands up towards his collar. Join his thumbs together and use a piece of *Motoyui*[41] string to tie them together as shown in the illustration. Pierce a hole in the back of his collar and thread the ends of the *Motoyui* cord through it and tie. Your prisoner will not be able to pull his fingers free. Due to the pain, he will also be unable to shake free.

If you do not have any Motoyui, then the string from a Shamisen, Japanese lute, can also be used.

---

[41] Motoyui is a special type of stiff string used to tie up the topknot and other parts of Japanese hair. The string was readily available due to common use. A piece was generally 28 inches/ 70 cm in length.

九字秘法

抑此九字ハ大摩梨支尊祕授兵法九字、大東ハ
身心堅固ニシテ運カヲ増ケ惡魔ヲ拂ヒ惡靈邪
鬼狐狸妖怪ヲ滅シ惣シテ一切ノ厄難ヲ除キ諸ノ願望ヲ
成就圓滿ナラシムル神術ナリ
至心ニ傳授シテクシク修スル人ニ靈驗著シキ也云々

切紙九字之大事

臨 りん

左右の手をくゝそ中指とたてあハん

獨古印 とこのいん

鬪 とう

外師子印 げーのいん

左右たびお小中指にて頭指とらくて大指死各指小指とさてあハん

兵 ひやう

大金剛輪印 だいごんごうりんのん

二手内ふくミ頭指とたて付中指かてくらむ

者 しや

内獅子印 おいしのん

左右さひ小中指ふて死名指とらくミ大指頭指小指とさてあハん

皆（かい）

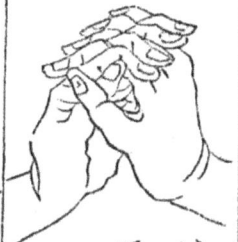

外縛印（げばくのいん）
二手おのく外へ
くミなり

烈（きつ）

智拳印（ちけんのいん）
右頭指とにぎり頭
指とそ左
のごとく右頭指とる

前（ぜん）

隠形印（おんぎょうのいん）
左の手とろろにぎり
右の手上におく呪恐ア剛
状おんてきこぶさん七難速
滅七福速生秘いきと分
ときえべ

日天子真言
おんでんしーしんごん
ち　ちゃ　くろ
摩利支天真言
おんまりしてんしんごん

陣（ちん）

内縛印（ないばくのいん）
十指たらひ
うちくミいを
なり

在（ざい）

日輪印（にちりんのいん）
左右の大指頭指の
端をほげ余四指ひ
らき散り

通六開四在八
者大者丼陣
兵元九
臨一闘三皆五烈七前九
天怪爛神力

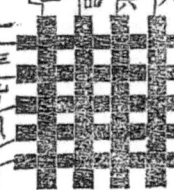

力印
左手ハ鞘
腰辺
おさ
を
おさ
く

九字ときる
ときめう
らう
くあり

Kuji Hiho Secret Nine Seals Technique

Originally Kuji,[42] the Nine Seals, was a secret teaching handed down to us by Marishiten.[43] *Heiho Kuji*, the Nine Seals for Martial Warriors, enables you to harden yourself both physically and mentally, while granting success in battle. At the same time, it drives away your hated enemies and banishes demons. It will completely rid your surroundings of evil spirits, malicious creatures, foxes, tanukis and ghosts. This divine technique severs any connection between you and natural disasters and allows all your desires to come to fruition.

Those that devote extensive time and energy to ingraining this into their being will find they are in possession of supernatural abilities.

---

[42] The seals are a set of certain Kanji paired with hand positions, that when implemented in a set form a protective spell for the user.

[43] Marishiten was originally a Hindu God but was later adopted by Buddhism as a representation of the wife of the sun god. Though often considered female, Marishiten can also be depicted as male.

Kirigami Kuji no Daiji The Nine Seals – Cut of Paper Level[44]

1. *Rin* 臨 – **Facing Off Against**
独古印 **Dokko Seal**[45]

Join your hands together, entwining your fingers together except for your middle fingers which should be extended with the fingertips touching.

---

[44] A Kirigami is a document certifying transmission of a certain level of initiation. Meaning literally "a small piece of paper cut off a larger piece" it refers to a basic document certifying proficient in a portion of an art.

[45] Dokko is also known as a vajra, which is a metal implement used in Buddhist rituals. The Dokko symbolizes both the hardness of a diamond and the destructive power of a thunderbolt.

独鈷杵Dokko-sho – One pointed Vajra

三鈷杵Sanko-sho – Three pointed Vajra

五鈷杵 Goko-sho – Five pointed Vajra

金剛杵

独鈷杵　　　　三鈷杵　　　　五鈷杵

兵

2. *Hyo*兵 – Soldier
大金剛輪印 **Great Golden Wheel of Strength Seal**[46]

Interlace the fingers of your right and left hands. Extend your index fingers and join the tips. Wrap your middle fingers over your index fingers.

鬪

3. *Toh* 鬪– Fight
外師子印 *Gejishino In* – **Outer Lion Seal**[47]

Wrap both middle fingers over both index fingers. Your thumbs, ring and little fingers should be extended with the tips touching.

---

[46] This seal allows you to dispel obstructions and evil.

[47] Your hands depict a lion. Your thumbs are ears, your index fingers are the eyes and the interval between the little and ring fingers are the mouth.

Side view	Top View

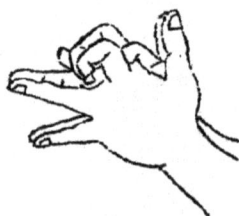

4. *Sha*者 – Person
内師子印 Inner Lion Seal

Bring your hands together so that your middle fingers intertwine your ring fingers. Extend your index and little fingers and touch the tips together. [48]

---

48   Side View   Top View

5. *Kai* 皆– **Everything and Everyone**
外縛印 **Outer Binding Seal**
Bring both hands together so that your fingers cross and extend across the backs of your hands.

6. *Jin* 陣– **Legion Organized For Battle**
内縛印 **Inner Binding Seal**
Interlace all ten fingers.

7. *Retsu* 烈– Split Apart
智拳印 Fist of Wisdom Seal

Extend your right index finger and squeeze the remaining four into a fist. Use your left hand to hold the tip of your right index finger as shown.

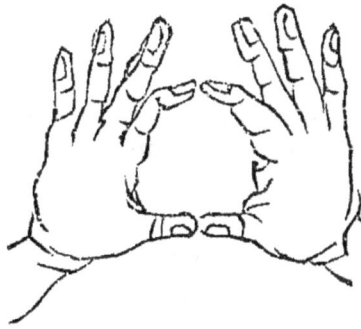

8. *Zai* 在– Exist
日輪印 Ring of the Sun Seal

Connect the tips of your left and right index fingers and thumbs. The other four fingers on each hand are spread out.

前

9. *Zen* 前 – In Front
隠形印 Hidden Shape Seal

Close your left hand into a fist, leaving a hollow space. Place your left hand on top of your right palm. There is an orally transmitted lesson regarding this.[49]

This will cause any demons plaguing you to be subdued, hated enemies to be repelled, will purge the seven misfortunes and bless you with the secret to being blessed with the seven types of good fortune.[50] You should be chanting this as you do Kuji Kiri, Cut the Nine Seals.

---

[49] Other versions of the Kuji give the orally transmitted instruction, which is: Join the fingernails of your left thumb and index finger with your right thumbnail. This is shown in the illustration.

[50] The final part of this spell contains several expressions:
Cause demons to submit:

Akuma Kofuku 悪魔降伏
Repel a hated enemy:

Onteki Taisan 怨敵退散
Instantly purge the 7 misfortunes:

Shichinan Sokumetsu 七難速滅. According to the Lotus Sutra, the seven misfortunes are fire, wind, floods, wars, punishments, demons, and thieves. Salvation can be found by reciting the name of Kannon, goddess of mercy.
Secret to instantly activate all seven types of good fortune:

Shichifuku Sokushohi 七福速生秘. Wealth, treasure, good fortune, good health, longevity, contentment and love.

Kuden
Oral Transmission for Zen – In Front

One version of this orally transmitted lesson is as follows.
Chant:
Akuma goufuku. Onteki taisan. Shichinan sokumetsu. Shichinan fuku sokuseihi (or fuku sokusho)

Down with devils,
Beat away enemies,
Quickly destroy seven troubles,
Recover strength seven times

日天・摩利支天
Nitten · Marishiten
Prayers invoking the Deities Nitten and Marishiten[51]

Edo Era woodblock prints of Nitten (center) and Marishiten (right.)

Nitten Jishigon
Mantra of Surya, one of the Twelve Devas guardian deities.[52]
On Adichi Yaya Sowaka – Send me your protection

Marishiten Shingon
Mantra of Marishiten[53]
Onmari Shiya Sowaka – Holy Marishiten I Praise You

---

[51] These prayers are written with Sanskrit characters with Japanese Hiragana alphabet indicating how they should be read.

[52] The Twelve Devas represent Indian gods that preside over the twelve direction-north, south, east, and west, the four semi-directions, up and down, and the sun and the moon.

[53] Marishiten was originally a Hindu God but was later adopted by Buddhism as a representation of the wife of the sun god. Though often considered female, Marishiten can also be male.

To-In
The Sword Seal

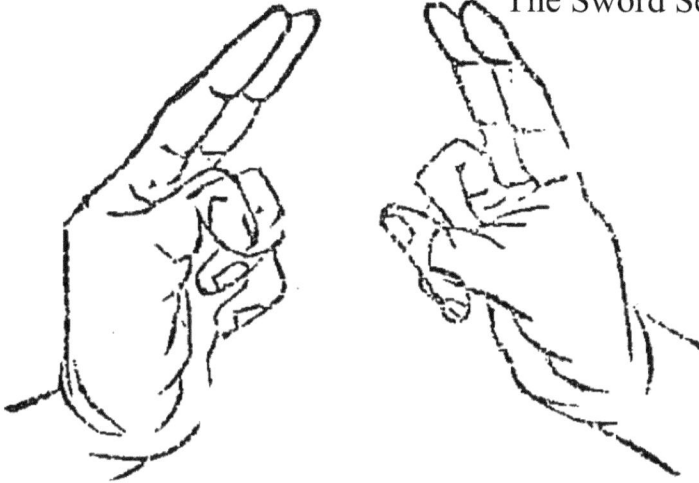

To make the sword seal, extend the first two fingers of each hand as shown in the illustration. Your left hand represents the scabbard and your right hand the sword. Move your hands to your hip and draw the fingers of your right hand out. Then cut the Kuji in the order shown with the two fingers of your right hand representing a sword while chanting.[54]

---

[54] Illustration showing how the fingers of your right hand are in the scabbard formed by the fingers of your left hand.
From :*Kuji for Self-Defense*
1812

Chant the following,

Rin, Pyo, Toh, Sha , Kai, Jin, Retsu, Zai, Zen
臨　兵　鬪　者　皆　陣　烈　在　前
Face, Soldier, Fight, Person, Everything, Legion, Arrange, Exist, Before

Celestial soldiers, descend and arrange yourselves before me![55]

			② 兵 Hyo	④ 者 Sha	⑥ 陣 Jin	⑧ 在 Zai
①	臨	Rin				
③	鬪	Toh				
⑤	皆	Kai				
⑦	裂	Retsu				
⑨	前	Zen				

---

[55] The English translation of each Kanji in the Kuji is approximate as the meaning of each Kanji can change according to the incantation being employed or by the school.

This phrase oringates in a book called The *Baopuzi* 抱朴子 or *The Master Who Embraces Simplicity*. It was written in 317 AD by the Taoist practitioner, philosopher, physician and politician Ge Hong (283~343 AD.) It is divided into twenty *Inner Chapters* and Fifty- two *Outer Chapters*. The inner chapters deal with Senjutsu, or Taoist Immortal Techniques, such as longevity and immortality. The *Outer Chapters* is a Confucian political treatise that discusses the merits and demerits of politics, as well as good and evil in human affairs. Thus the book is commonly known as *The Inner and Outer Books of Taoism.*

This phrase simultaneously means,

Sore, Gen, Kai, Dai, So, Ben, Jin, Tsu, Riki
夫　元　怪　大　燥　弁　神　通　力
That Person, Origin, Bewitched, Great, Fan Flames, State, Divine, Trancendental, Power[56]

			② 元 Gen	④ 大 Dai	⑥ 弁 Ben	⑧ 通 Tsu
①	夫	Sore				
③	怪	Kai				
⑤	燥	So				
⑦	神	Jin				
⑨	力	Riki				

---

[56] The author does not give any further information on how these two phrases interact.

破軍星線法

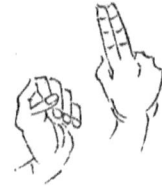

刀印ヲ結ビ九字ヲ唱ヘガラ図ノ如ク切ルベシ但シ其事ニ随ヒ顔事ヲ唱ヘ上図ノ左ニ握リタル大

指食指ハ間ヘ口ヲ当テ息ヲ吹掛ケ直チニ刀印テ押ヘ付左リ腰ニ付前ノ如々向ヘ切付ルナリ

書シ十字大指ニテ握リ行ナリ

ノ行事ニ随ヒ男ハ左掌ニ刀印ニテ

此字毎ニ註解アリ其ハ身

十字大法ノ解

月	ツメ
一月	五ツメ
二月	六ツメ
三月	七ツメ
四月	八ツメ
五月	九ツメ
六月	十ツメ
七月	十一ツメ
八月	十二ツメ
九月	一ツメ
十月	二ツメ
十一月	三ツメ
十二月	四ツメ

一月五ツメと八一月子のとき

十字大法　文字

天 航海渡	王 裁判ト	神 悪人招行人へ	日 病者ヲ訪フキ
龍	命 敵ニ向フトキ	大 悪爽見舞ノトキ	
虎 山猟或賊ニ應ル	勝 賣買ノトキ	水 水件ニ行トキ	

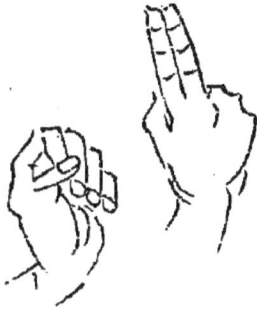

嵐吹く 外山の霞 曇無く 向ふ悪事を 皆切り拂ふ

Fierce storms do not blow,
Neither mist nor clouds surround the mountain near the village,
Sweep away all evil before me[57]

To use the Nine Seals, the fingers of your right hand form a sword, which you cut with. However, to ensure the spell fulfills your prayer, make a fist with your left hand as shown in the illustration.

Then place your mouth against the opening made by your index finger and thumb and blow the chant through your hand. Then, immediately sheathe the sword fingers of your right hand in the sheath of your left hand and bring both hands down to your left hip. After doing this, draw the sword and cut the Nine Seals as previously described.

---

[57] The cursive text has been rendered into typeface and then rendered in contemporary Japanese.

An Explanation of Juji, the Tenth Seal

There is an explanation for each seal and you should use the appropriate one based on the situation. Men should draw the Kanji on the palm of the left hand using Sword Seal. Make sure to grip your thumb as you do this.

- Ten 天 Heaven

Use Heaven when crossing a body of water by boat, or when your opponent is higher status.

- Ryu 龍 Dragon

Use Dragon when setting out on a journey into heavy wind or rain.

- Tora 虎 Tiger

Use Tiger when hunting in the mountains or to protect yourself when faced with bandits.

- Ou 王 King

Use King when you have to suddenly make a critical decision.

- Inochi 命 Life

Use life when facing an enemy.

- Sho 勝 Victory

Use Victory when buying or selling anything.

- Kami 神 Divine Being/God

Use Divine when going to a place where violent treacherous men reside.

- Dai 大 Great

Use Great when visiting a person who has the plague.

- Sui 水 Water

Use water for anything to do with water.

- Hi 日 Day

Use Day when visiting a sick person.

How to Control *Hagunsei*, the Army Breaking Star

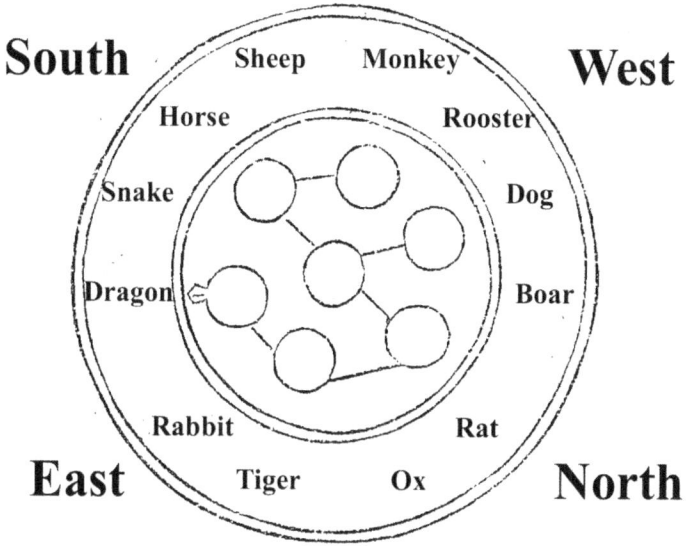

South **West**

Sheep   Monkey

Horse   Rooster

Snake   Dog

Dragon   Boar

Rabbit   Rat

East   **North**

Tiger   Ox

Translator's Note: An illustration of the Army Breaking Star depicted as a Dragon

From : *A Guide to Military Ceremonies* : Mist Volume 軍礼（霜）
Author unknown, Late Edo

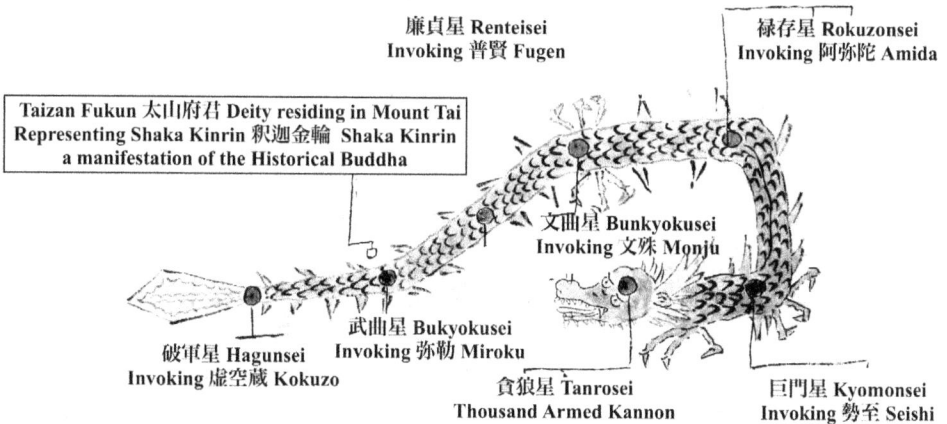

廉貞星 Renteisei
Invoking 普賢 Fugen

禄存星 Rokuzonsei
Invoking 阿弥陀 Amida

Taizan Fukun 太山府君 Deity residing in Mount Tai
Representing Shaka Kinrin 釈迦金輪 Shaka Kinrin
a manifestation of the Historical Buddha

文曲星 Bunkyokusei
Invoking 文殊 Monju

破軍星 Hagunsei
Invoking 虚空蔵 Kokuzo

武曲星 Bukyokusei
Invoking 弥勒 Miroku

貪狼星 Tanrosei
Thousand Armed Kannon

巨門星 Kyomonsei
Invoking 勢至 Seishi

How to Control *Hagunsei*, the Army Breaking Star[58] which is part of the Big Dipper[59]

If it is January, the first month,[60] and your duel will be at the hour of the Rat, 11pm~1 am, then you would count clockwise five intervals from Rat, meaning you end up at Dragon. This will mean your back is to the Army Breaking Star, the strongest and most auspicious position. Your sword should face that direction when you duel. The following chart shows how to calculate your position. [61]

---

[58] When considering a battle plan, one of the first steps that a Warring States Period Samurai general would take would be to find which direction the star known as *Hagunsei* 破軍星 or "the army breaking star" is facing. *Hagunsei* is the star at the end of the handle in the Big Dipper. Also known as the *Kensakisei* 剣先星 tip of the sword star.

[59] In Chinese tradition, the Big Dipper is made up of seven stars and as a whole represents *Taizan Fukun* 太山府君 the deity who resides in Mount Tai in China and serves as one of the Ten Kings of Hell under Tenma, the lord of hell. It represents 釈迦金輪 *Shaka Kinrin* 釈迦金輪 Gold Wheel Shaka. *Shaka Kinrin* is a manifestation of Shaka Buddha 釈迦如来 the Historical Buddha. The other stars are as follows:

貪狼星 Tanrosei	Invoking 千手	1000 Arm Kannon
巨門星 Kyomonsei	Invoking 勢至	Seishi
禄存星 Rokuzonsei	Invoking 阿弥陀	Amida
文曲星 Bunkyokusei	Invoking 文殊	Monju
廉貞星 Renteisei	Invoking 普賢	Fugen
武曲星 Bukyokusei	Invoking 弥勒	Miroku
破軍星 Hagunsei	Invoking 虚空蔵	Kokuzo

[60] This is referring to the old calender used before adopting the western calender. The Western months are given as a refrence.

[61] In Onmyoji Japanese sorcery, the final star in the handle of the big dipper is considered to be the tip of a sword and used to determine good or bad fortune. In the Asuka period (550-710 AD) Onmyoji combined Taoist, Shinto and Buddhist beliefs and established new rituals.

Sign	Represents the month	Number of spaces to count	Time
Ne 子 Rat	November	3	11 pm ~1am
Ushi 丑 Ox	December	4	1~3am
Tora 寅 Tiger	January	5	3~5 am
U 卯 Rabbit	February	6	5~7 am
Tatsu 辰 Dragon	March	7	7~9 am
Mi 巳 Snake	April	8	9~11 am
Uma 午 Horse	May	9	11am~1pm
Hitsuji 未 Sheep	June	10	1~3 pm
Saru 申 Monkey	July	11	3~5 pm
Tori 酉 Rooster	August	12	5~7 pm
Inu 戌 Dog	September	1	7~9 pm
I 亥 Boar	October	2	9~11 pm

年中風雨考（ねんぢうふううかんがへ）

五六月のころ風南まハれバ必ようあり
十一十二月
正月の頃風北へまハれバ日よりあり
五月西春ハ南ヨ秋ハ北いつルとちヨて雨ふると志れ

夜九ツよりふりだせハおかのあむ
明け七ツハひるまでやむ
あさ五ツハおちかぬあり
ひる九ツハむらくまでやむ
ひる七ツハおりくあめあり
ひる五ツハだらりくまでやむ
夜八ツよりふりだせハほどなく止
明け六ツハ半日ふるあり
ひる四ツハやがてはるゝあり
ひる八ツ半日ひとふるあり
くれ六ツハやがてはるゝあり
夜四ツハ半夜ひとふるあり

右大暦ナリス夕立と時雨ハ考へあるべからず

成就日（じやうじゆにち）

正月	二月	三月	四月	五月	六月	七月	八月	九月	十月	十一月	十二月
とら	み	さる	い	う	むま	とり	ね	たつ	ひつじ	いぬ	うし

不成就日（ふじやうじゆにち）

正月 七月	五日 十一日 十九日 廿七日
二月 八月	二日 九日 十六日 廿六日
三月 九月	朔日 九日 十七日 廿五日
四月 十月	四日 十二日 廿日 廿八日
五月 十一月	五日 十三日 廿一日 廿九日
六月 十二月	六日 十四日 廿二日 晦日

How to Interpret the Wind and Rain Throughout the Year

In the fifth and sixth months of the year, if the wind starts coming out of the south, then the weather will be clear.

If the wind starts coming out of the north in the eleventh month, twelfth month or around New Year's, then the weather will be fine.

Know that if the wind is out of the west in the fifth month, out of the south in spring, out of the north in fall, then the spring winds from the east will bring rain.

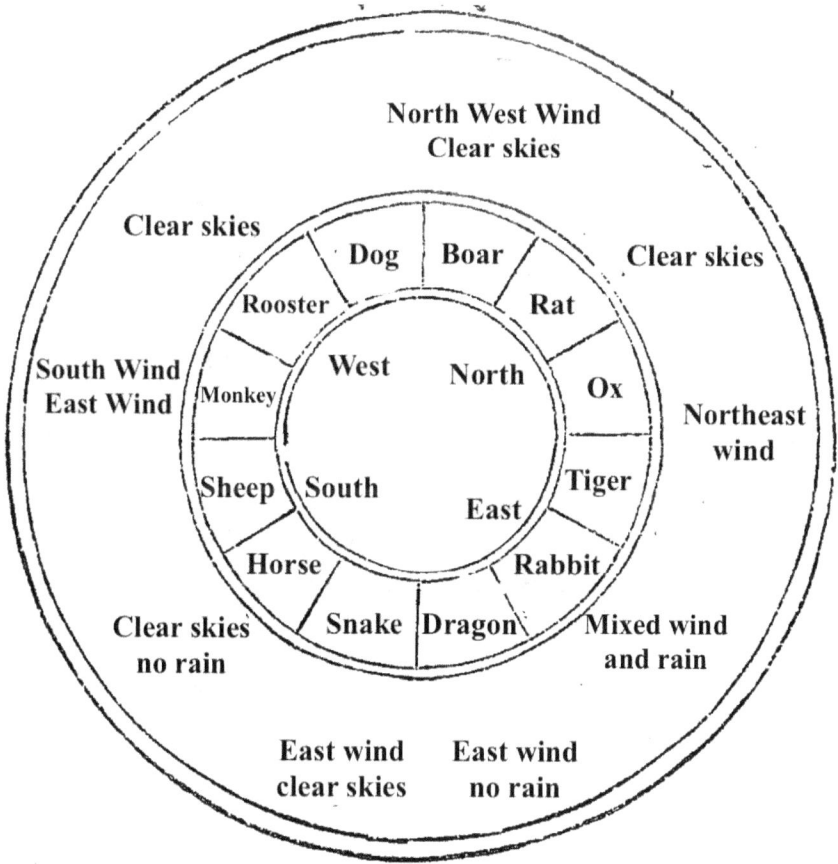

How to Interpret the Wind and Rain Throughout the Year

- If rain falls during the ninth hour of night, 11pm~1am, then it will be light rain.
- If rain falls during the eighth hour of night, 1am~3am, then it will soon cease.
- However, if rain begins falling in the seventh hour of night, 3am~5am, then it will persist until gradually stopping.
- Rain that begins in the sixth hour of dawn, 5am~7am, will likely continue for half a day.
- If it starts to rain during the fifth hour of dawn, 7am~9am, then it will be a long rain.
- Rain in the during the fourth hour of daytime, 9am~11am, means it will stop and then clear.
- Rain that begins during the ninth hour of daytime, 11am~1pm, then it will persist until gradually stopping.
- Rain that begins during the eighth hour of daytime, 1pm~3pm, like likely continue for half a day.
- Rain that begins during the seventh hour of daytime, 3pm~5pm, like likely continue for a long time.
- If it begins to rain during the sixth hour of evening, 5pm~7pm, then it will likely stop and soon clear.
- Rain that begins during the fifth hour of evening, 7pm~9pm, will likely persist for a while until gradually stopping.
- Rain that begins during the fourth hour of night, 9pm~11pm, will likely be a hard rain.

This is a general overview but one cannot forget to consider the possibility of *Yudachi*, sudden, heavy showers on summer afternoons or evenings as well as *Shigure*, rain showers in late autumn and early winter.

Jojubi
Lucky Days

January:	Day of the Tiger
February:	Day of the Snake
March:	Day of the Monkey
April:	Day of the Boar
May:	Day of the Rabbit
June:	Day of the Horse
July:	Day of the Rooster
August:	Day of the Rat
September:	Day of the Dragon
October:	Day of the Sheep
November:	Day of the Dog
December:	Day of the Ox

Fujojubi
Unlucky Days

January and July :	The 5th, 11th, 19th and 27th
February and August :	The 4th, 9th 16th and 26th
March and September:	The 1st, 9th, 17th and 25th
April and October:	The 4th, 12th , 20th and 28th
May and November:	The 5th, 13th, 21st and 29th
June and December:	The 6th, 14th, 22nd and last day of the month.

予ガ家世々武技ヲ以テ天下ニ鳴ル予生ヲ舊新ノ交ニ
得祖先來ノ遺傳ヲ受ケ聊カ此道ノ天下ノ利器タルヲ
知リ少壯ニシテ父ニ隨ヒ四方ニ周遊シ苟モ武ヲ以テ
名アルノ士邊偶地ト雖モ訪問セザルナシ如此困苦ヲ
積ミ辛酸ヲ嘗メ竟ニ此道ノ奥妙ヲ窺フヲ得クリ今ヤ
文弱ニ流ル、時此道ノ全廢センフヲ恐レ于此予カ
自得スルモノヲ圖解ヲ以テ後世ニ傳ヘントシ竟ニ此
一小冊子ヲ成セリ然ルニ發賣ノ初ヨリ國ノ內外ヲ問
ハス陸續購求セラル、ノ幸ヲ得テ予ガ面目是ニ過ギズ
然ルニ世儼悸私利ニ汲々タルノ徒アリ此書發賣ノ多
キヲ羨ミ窃ニ類似ノ書ヲ著シ又ハ字句圖解ヲ少變若

Conclusion

For generation after generation my household has taught martial arts. I was born during the change from the feudal Edo Era to the modern Meiji Era and I too seek to pass on to the next generation, some small amount of the teachings they have passed on to me from my ancestors. As a youth I followed my father's command to travel to all corners of the nation, visiting famous instructors. Over time and with great effort I was able to gain an understanding of the inner mysteries of classical Japanese martial arts.

Recently, with the tendency of people to devote more and more of their leisure time to reading books and magazines, I fear that martial arts training is beginning to disappear. In response to this I have created this illustrated guide in order to transmit this knowledge to later generations.

Even before sale of this book began, I was blessed to receive inquiries, one after the other, from both inside Japan and from foreign countries as well, asking where it could be purchased.

クハ少加シテ飜刻ヲ試ミントスルモノアリトカヽル
所爲ハ固ト國法ノ許サヾル所ト雖モ又其眞ヲ失シ後
世ヲ誤ルニ至ル豈ニ少シモ假借ス可キモノナランヤ
後來若シカヽル輩ヲ聞知シ玉フ諸君ハ此道ノ爲メ幸
ニ忠告アランコヲ希フナリ

著者兼
出版人

敬 白

While I am honored at such an enthusiastic reception and many readers are studying it diligently in order to improve themselves, there are some who are jealous. They are producing the same book with slight alterations or other small changes.

This type of behavior is a clear violation of the laws of our country and will result in incorrect information being transmitted to later generations. This activity should be stopped immediately.

For anyone who desires to preserve this way for future generations, please report any such activity to us.

Sincerely Yours,
The Editor and Publisher

Published November 21st 1887[62]

---

[62] By 1900 this book was on its 21st edition.